Swimming Against the Tide

African American Girls and Science Education

Sandra L. Hanson

 TEMPLE UNIVERSITY PRESS
Philadelphia

Temple University Press
1601 North Broad Street
Philadelphia PA 19122
www.temple.edu/tempress

Copyright © 2009 by Temple University
All rights reserved
Published 2009
Printed in the United States of America

Library of Congress Cataloging-in-Publication Data

Hanson, Sandra L.
    Swimming against the tide : African American girls and science education / Sandra L. Hanson.
        p. cm.
    Includes bibliographical references and index.
    ISBN 978-1-59213-621-6 (cloth : alk. paper)
    ISBN 978-1-59213-622-3 (paper : alk. paper)
        1. Science—Study and teaching—United States. 2. African American girls—Education—United States. 3. Sexism in science—United States. 4. Racism in education—United States. I. Title.

Q183.3.A1H367 2009
500.82—dc22   2008014330

2 4 6 8 9 7 5 3 1

Swimming Against the Tide

To Dorothy and Robert Hanson, with love.
You made me believe that anything was possible.

 Contents

Preface and Acknowledgments

The president of one of the most elite universities in the United States—Harvard University—addressed a group of economists in 2005 (Summers, 2005). In his comments, Lawrence Summers referred to an innate difference in science ability favoring boys. Those of us who have been studying gender and science over the past few decades were chilled by these comments. A large group of scientists have provided overwhelming empirical evidence that young boys and girls start out with very similar interests and abilities in science. School systems and a culture of femininity in the United States and elsewhere work to systematically encourage this interest and ability in boys and discourage it in girls. And so, young girls lose their interest and confidence in science as they continue on in the education system. This "cooling out" process results in a scientific labor force that is largely male. In this book, I examine a group of young women who I found show considerable interest and involvement in science—African American women. The U.S. science system is not just a male system. It is a white male system. Why, then, are these young women interested in science? Do they experience a double disadvantage from being female and black? The answers to these questions will help us understand the strengths and abilities that sometimes exist among groups of young people whose statuses (or multiple statuses) can be expected to universally work against them. My research supports the work of social scientists who insist on using a multicultural lens. It also supports those innovative education scholars who look at people's responses to social structures and not just the social structures themselves. Sometimes, just sometimes, young people do not absorb all of the education system's gender and race messages and obediently comply. Thus, this book is a work of hope and optimism in the context of continued racism and sexism in science. In the words of the young African American women who were surveyed, it explains just how they "swim against the tide."

Acknowledgments

This research was funded by a REC-0208146 grant from the Division of Research, Evaluation, and Communication of the National Science Foundation. The opinions expressed here do not necessarily reflect the position of the National Science Foundation. The author thanks Michelle Jiles, Yu Meng, Melissa Cidade, and Gail Smith for their analytic support.

Introduction

Understanding Young African American Women's Experiences in Science

T he study of elites has historically been an important part of social-science theory and research. Elites have been described as those occupying powerful and influential positions in government, corporations, and the military. These elites share interests and attitudes, and have networks that work to encourage and include some but discourage and exclude others (Domhoff, 1983; Mills, 1956; Zweigenhaft and Domhoff, 1998). In a technologically advanced, postmodern, global society, the status, power, shared interests, and powerful networks of those in science suggest that they must be considered as members of the new elite. One of the most distinguishing features of the science elite (historically and currently) is the shortage of women and non-whites. In spite of the progress that women and minorities have made in science education and occupations (Hanson et al., 2004; National Science Foundation, 2000, 2004), the culture of science continues to be a white male culture that is often hostile to women and minorities (Catalyst, Inc., 1992; Harding, 1986; Rossiter, 1982, National Science Foundation, 2000, 2004; Pearson and Bechtel, 1999; Ramirez and Wotipka, 2001).

Although research on women in science has proliferated, its focus has often been on differences between men and women, with little attention to subgroups of women. It is a mistake to think of women as an undifferentiated group. Increasingly, researchers have come to the conclusion that not all women have the same experiences in science education and occupations (Hanson and Palmer-Johnson, 2000; Mau et al., 1995). In fact, preliminary research has suggested that young African American women are particularly interested in science (sometimes more so than their white counterparts) (Hanson and Palmer-Johnson, 2000; Hanson, 2004; National Center for Education Statistics, 2000a). In spite of this interest, some have suggested that African American women experience racism and sexism in the science domain and remain underrepresented in science programs and occupations (Malcom et al., 1998; Vining Brown, 1994; National Science Board, 2000). The paradox at the center of this book involves the expressed scientific interest and motivation of many young African American women in the context of a science culture that remains openly hostile to those who are not white and not male.[1]

The conceptual framework that is used here in trying to understand the complex interaction between race and gender in the science domain is one that focuses on gender and race as structures that are major principles of organization and inequality.[2] It is not qualifications but master statuses such as gender and race that are viewed as being important determinants of entry and success in science. This framework is multicultural and feminist in suggesting that gender structures are powerful aspects of organization but they are not identical across cultures. It is important that diversity in gender systems across race/ethnic groups is acknowledged. The framework also borrows from the critical perspective in suggesting that it is not structure alone that creates and discourages opportunities in science. Women and African Americans and African American women are not merely "victims" of racism and sexism. Sometimes structures and status quos are questioned and challenged. Thus, it is not just structure that is important for understanding African American women's experiences in science but these young women's responses to race and gender structures.

My goal in this book is to examine the experiences of young African American women in science education. First, I describe these young women's attitudes about science, their course-taking in the science curriculum, and their science achievement. Additionally, I report on the extent to which these young women see discrimination and a chilly climate in the science classroom. A second goal of the book involves an examination of young African American women's experiences in school systems, families, communities, and peer groups that work to encourage or discourage their interest and achievement in science. The major research questions revolve, then, around the experiences of young African American women in science and the push-and-pull factors in a variety of areas of life that influence these experiences. My special focus is on experiences in science education during the high school years, although later experiences are also considered. In order to understand the way in which race and gender come together, comparisons with young white women and young African American men are included.

The research questions are addressed using a triangulation of methods and data sets. A chapter is devoted to each area of influence: school, family, community, and peer. The chapters begin with a short discussion of relevant literature. Analyses of the National Educational Longitudinal Study (NELS) of American youth are then presented. These analyses lend insight into the experiences of a large, nationally representative sample of high school students in the science curriculum. The data include rich information on science experiences, as well as experiences involving schools, families, communities, and peers, thus allowing measurement and understanding of the processes at work in young African American women's science experiences.

The examination of NELS data will be supplemented by information from a Web survey of young African American women that I conducted in 2003. The Web survey attempted to gain further insight into young African women's science experiences by using vignettes and open-ended qualitative questions. The survey used a new Web technology that combined probability sampling with the reach and capabilities of the Internet. The open-ended questions (as well as focused, closed-ended questions) inquired specifically about the variety of influences in family, community, peer groups, and schools that might have encouraged or discouraged these young women in science education. As Feagin, Vera, and Imani (1996) suggest, it is critical that we allow minorities to explain their experiences in their own words.

Additionally, my Web survey used the vignette technique to gauge young African American women's perceptions of discrimination in the science classroom. I know of no other researchers who have used the vignette methodology in the context of the Web survey. The attitude survey has become standard technique in social science research. However, it can be argued that questionnaires and surveys are not well suited for the study of human attitudes and behaviors since they often produce unreliable and biased self-reports (Alexander and Becker, 1978). Standardized surveys often use vague questions that are interpreted within the respondent's own meaning system. One solution, then, is to make the stimulus given to the respondent as concrete and detailed as possible, as in a vignette. This stimulus should resemble a real-life decision making or judgment-making situation. In addition, since the stimulus can be held constant or varied in a systematic fashion, it gives the researcher greater control, as in experimental designs. The vignette is an ideal strategy for gaining this detail and control.

The vignette in the Web survey began with the respondent being shown a picture of a young woman. One part of the sample was given a picture of a young African American woman, LaToya. The respondent clicks on the picture and listens to the following vignette, provided in LaToya's voice:

> Hi, my name is LaToya. I'm an African American student and I go to high school in Northern Virginia. I love science. Especially biology. I am thinking of maybe becoming a veterinarian. But even though I love science, I don't feel so welcome there. I mean like the other day I was sitting in my biology class and all of a sudden I noticed that I don't get called on very often in this class. And it's like my teacher's white and most of the students in the class are white. Then I look in my textbook and see pictures of all of these white scientists. Sometimes I don't feel like I belong in science.

This vignette was followed by a series of questions that attempted to measure specific responses, experiences, feelings, and sources of encouragement and discouragement in science. Most questions were qualitative and some, but not all, refer to the vignette. Distinction was made between their own experiences and those of minority women in general. Some questions asked for specific examples of encouragement and discouragement from specific sources (e.g., teachers, parents) and other questions asked whether the respondent thinks that their abilities and interests are influenced by feedback from these sources. Finally, they were asked what they would do to change the school system so that all students would feel encouraged in science. Note that other forms of the vignette involved LaToya talking about gender as a problem in the classroom. In another vignette, the respondent listened to a young white woman, Michelle, talking about gender discrimination in the science classroom. A fourth, neutral vignette is also given where LaToya feels unwelcome but does not mention race or gender. Thus, the race of the target was varied as was the content of the vignette (focusing on race or gender [or neither] in the classroom). Keep in mind that each of the vignettes was given to a sample of white women and each was given to a sample of African American women. Respondents only received one version of the vignette.

Race discrimination is a sensitive topic and standardized questions have the potential for generating socially desirable responses. The vignette provides a solution. It avoids asking about the respondent's own experiences but rather presents a hypothetical situation. Additionally, the Web survey allows a greater feeling of anonymity, again contributing to more honest, valid responses.

The paradox of young African American women's interest in science in the context of a chilly climate for those who are not white and not male is a complex one that demands a complex explanation. The literature and analyses presented here within the context of the critical feminist framework present a picture of a school system that does not tend to see young African American women as science talent. That is not to say that there are no teachers who make a difference for these young women. Overall, however, the science education system is structured in a way that favors white middle-class males. The answer to the paradox continues to evolve as the African American family, community, and peer systems are examined. As research and theory on human agency suggests, sometimes events and circumstances come together in unexpected ways. The African American family and community have a different definition of femininity than does the white family and community. African American women have historically been viewed as strong individuals who work and head families. Gender is often thought of in a universal way—especially within a particular society. I provide evidence here of a unique gender system in the African American family, community, and peer systems that sometimes works against all odds to encourage interest and activity in science. The above statements are phrased at a very simplistic level. The actual processes revealed by the NELS and Web survey data allow the reader to see the complexity and subtleties of the push-and-pull factors that influence young African American women in the science domain. It is when these young women are allowed to provide answers in their own words—in the Web survey—that the paradox of young African American women in science is unraveled.

This research examines gender structures in science and in African American communities. More specifically, it attempts an understanding of young African American women's experiences in science and the factors that encourage and discourage success there. Data from a variety of sources is used to chronicle these experiences and share the young women's view of them. When gender and skin color are the major factors determining who will do science, a considerable amount of scientific talent is lost. The implications of this talent loss for scientific discovery and advance are considerable. The implications are also great for the young people who are denied access to science since they will not be involved in the creation of policies and technologies that will guide us through the next century.

2 The Conceptual Framework

A Critical-Feminist Approach

The conceptual framework used here acknowledges the patriarchal nature of the science system and the power of gender and race structures in creating and limiting opportunities for individuals. However, it is multicultural in its view of gender. It also avoids a deterministic, victim approach to looking at the experiences of young African American women in science education. I use a critical-feminist framework that acknowledges the dialectic between structure and agency.[1] In the pages below, I elaborate on the assumptions of the feminist framework and critical approach used here. I also argue the need for a particular multicultural feminist framework. Thus, my conceptual approach acknowledges both gender and race discrimination in science *and* agency and resistance in gender cultures promoted by African American families and communities.

Increasingly, research on gender has moved away from seeing gender as a learned trait and a property of individuals who are socialized into a particular gender role. Feminist theory has been instrumental in this movement. The gender role approach tended to make gender differences appear natural and did not question where the ideas and behaviors socialized into young boys and girls came from. More recent theorizing on gender (often, but not always, falling under Marxist and neo-Marxist feminist frameworks) takes a structural approach and argues that gender constitutes a macrostructure that hierarchically stratifies society. Thus, a study of gender relations is similar to a study of class relations or race relations, and the notion of sex roles is as inappropriate (in most conceptual frameworks) as is the notion of race roles or class roles. Gender is seen here as a principle of organization rather than a characteristic of individuals—it is something people do in their everyday lives, it is constructed at all levels of social life, and it is built into our social institutions. Within this organizational framework, the emphasis moves away from socialization and toward examination of processes that categorize and stratify. For example, Kantner's (1977) study of women in corporations did not focus on women and their socialization into unique attitudes about work but rather on work structures that create low attachment to (often female-dominated) low-end jobs. She found that whether male or female, people in certain jobs develop similar attitudes toward work. Thus, power is a central concept in structural theories and gender is viewed as something that is constructed (Lorber, 1994). The study of gender becomes a study of power hierarchies and the reproduction of these hierarchies.

Although macrotheories of gender make a contribution by recognizing gender as a structure based on power relations, they do not go far enough in acknowledging history, ideology, and, especially, agency in gender structures. Structural theories of gender often appear deterministic, failing to point out the fact that social structures such as gender can be re-constructed as well as constructed. A critical approach to gender avoids the determinism of many structural theories and allows for a dialectic between structure and agency. Just as structure conditions life, people change social conditions (Agger, 1998). Studies of gender and education that have taken a critical approach examine processes that challenge status quo gender structures, as well as processes that create gender inequality. For this reason, the critical perspective is sometimes referred to as resistance theory (Grant et al., 1994). Work by education scholars who take this approach argues that students do not absorb all messages (e.g., about gender, race, and class) and obediently comply with educational structures without protest (Grant et al., 1994; Giroux, 1983). It is not just structures (as emphasized by macrostructural theories) that are important for understanding education systems but also people's responses to those structures. Ultimately, the study of structures is also a study of personal choices and values, since these structures of gender relations enter into our everyday lives and work to shape our personalities as men and women.

In addition to its focus on agency, the critical approach tends to put more focus on ideology and history than do most structural gender theories. Recognition of the extent to which ideologies help reinforce structures is important. Powerful ideologies—reflected in dominant belief systems—help create a false consciousness that allows dominant groups to rule by consent and not force. Thus, gender hierarchies are maintained in part by dominant patriarchal belief systems that favor males and justify the gender order as a natural one (Sage, 1990). Most women willingly accept the ideologies and cooperate with them (Connell, 1987). In the following chapters, I discuss ideologies about science as a white male domain that are presented to young women in general and young minority women in particular. I also discuss unique ideologies about women, family, and work in the African American community—a key factor in women's expectations and behaviors regarding science.

This critical approach also puts a strong emphasis on historical context. The concept of agency is relevant to this historical perspective. It is not enough to assume that gender relations are produced and reproduced vis-à-vis power structures built into social systems. One must acknowledge that there is constant negotiation and maintenance involved in keeping hierarchies in place. Sometimes structures are questioned and historical and structural conditions combine to create a sense of agency for women, leading to victories that may or may not restructure the gender hierarchy and change the course of history, but do contest it (Connell, 1987). The combination of individual agency and historical consideration provide for elements of choice and transformation. One cannot consider any gender system without considering the historical development of that system. There is a long and unique history of women's combined work and family roles in the African American community (in part related to the experience of slavery), as well as a historical record of family investments in sons and daughters that are not replicated in white communities.

Thus, our critical approach will demonstrate how the African American family system, both *historically* and at present, displays a unique set of gender *ideologies* and interactions that might be a source of *agency* (and strong personalities) for young African American women despite contemporary (and historical) race and gender structures that work against them in the science domain.[2]

The particular feminist framework that I use, in conjunction with the critical approach, is a multicultural one that acknowledges patriarchal structures but also recognizes the diversity in gender systems and the white, middle-class bias of much work on gender. It derives from work by women of color who have attempted to correct these biases in social science research (Zinn and Dill, 1996; Collins, 1990a, 1999).

Here, the focus is not on gender alone but on the intersection between race, class and gender (Andersen and Collins, 1995; Rothenberg, 1992; Glenn, 1985; West and Fenstermaker, 1995). Collins (1990a, 1999) has been particularly effective in describing this approach.

It is ironic that early work on gender biases failed to see its own white, middle-class bias (Collins, 1999). Collins suggests that people can be oppressed and oppressor, privileged and penalized—simultaneously. No one form of oppression is primary. Rather, there are layers of oppression within individual, community, and institutional contexts. Like critical-gender theory, she argues that each of these locations is a potential site of resistance. The approach also emphasizes the unique history and subcultures of minority women and the role of these cultures in affecting both the structures that limit and direct women's lives as well as the agency that allows them to retain some control and influence within these structures (Hanson and Kraus, 1998, 1999; Collins, 1999). The multicultural approach suggested by Collins (1999) stresses the import of not only structures but also ideologies. Here, the focus is the unique subculture of African American women (and their families and communities) and its intersection with women's experiences in science education. As the literature in the chapters on African American families and communities suggests, African American subcultures might provide young women a unique set of resources that allow for and support interest in science among these young women. Although these families and communities may not always have high levels of economic resources, they compensate for disadvantages in some resources with an excess of other resources (e.g., unique gender ideologies, work expectations, and family expectations). This context is an important one for understanding young African American women's behaviors and personalities. Family characteristics often influence the science achievement outcome through the actions and attitudes of the individual (Hanson, 1996). For example, there is research suggesting that relative to young white women, young African American women show less fear of success (Murray and Mednick, 1977) and experience less gender stereotyping and fear of rejection by young minority men when considering a career (Gump, 1975; Scanzoni, 1975).

It should be noted that, in classifying this approach as feminist, I acknowledge, like feminist theory in general, 1) gender is a major source of stratification and differentiation; 2) gender is constructed and thus can be re-constructed; and 3) a commitment to changing inequitable gender arrangements. However, this framework does not fall simply within any of the usual feminist categories such as macrostructural, microstructural, or interactionist (Chafetz, 1999). In combining the critical approach with the feminist framework and making it multicultural, the best description of the framework would be critical-multicultural feminist. I do not suggest that my particular feminist approach is superior to others. In fact, reviews of feminist theory, such as that by Chafetz (1999), reveal the richness in variety of feminist theory and the necessity for this diversity, given the complexity of gender questions at all levels of interaction. However, given the particular issues and complexities involved in this research on young African American women in science, the approach I use here is most useful in answering the research questions.

3 Young African American Women's Experiences in Science

"Science Is Like Opening a Present from Your Favorite Aunt. You Just Can't Wait to Open It Because You Know There Is Something Wonderful and Unique Inside."

Before describing the factors in young African American women's lives that encourage and discourage them in science, it is important to describe these young women's science experiences involving attitudes toward science, course-taking in science, and science achievement. The quandary that this book addresses involves an interest and involvement in science among young African American women that exists in spite of the fact that their gender and skin color might be expected to put them in double jeopardy in the white male science culture. In the following pages, I first present some background from the literature on the topic of race, gender, and science achievement. I then analyze data from a large national data set—the National Educational Longitudinal Study (NELS)—to provide information on traditional measures of science education experience in a representative sample of young African American women. Throughout these discussions, comparisons are made across race (e.g., with white women) and gender (e.g., with African American men). I also use quantitative data from a Web survey of young African American women to provide more recent information on science achievement, access, and attitudes from a sample selected specifically for my research questions. Both quantitative and qualitative data from the Web survey are used to reveal the extent to which young African American women feel welcome in science.

The final aspect of the Web survey examined here involves the young women's responses to a series of vignettes. As described earlier, these vignettes involve a young woman (sometimes African American, sometimes white) talking about her experiences in the science classroom. The variation in sample (African American and white respondents), race of target, and content of vignette allow insight into the extent to which (and conditions under which) young African American women see discrimination in the science classroom.

Background: The Literature on African American Women's Science Experiences

Some have argued that since women perform less well in science than men, and minorities do less well than whites, there will be a double disadvantage for young African American women in science (Clewell and Anderson, 1991; Cobb, 1993; Vining Brown, 1994). The double jeopardy argument assumes an additive effect of the two statuses—being female and African American. The tendency in science research has been to look at women or minorities and not examine variation within these groups. Hence, there is little research on minority women in science (Burbridge, 1991; Catsambis, 1995). The additive argument does not acknowledge the fact that gender and race might interact to affect science outcomes. It is only by looking at science experiences among women from different race/ethnic groups, and across gender groups within African American populations, that a true understanding of the way race and gender come together to affect science experiences can be achieved. Double jeopardy, in terms of acceptance within the science system, may be found but that does not necessarily convert to double jeopardy in these young women's interest and involvement.

It is true that non-Asian minorities and women do less well in science when compared to whites and men, respectively (National Science Foundation, 1999, 2004). Some research has shown that African American women remain underrepresented and have relatively higher attrition rates in undergraduate and graduate science programs and in science occupations (National Science Board, 2000; National Science Foundation, 2000). There is evidence that young African American women experience special barriers in science (Carwell, 1977; Vining Brown, 1994; Clewell and Anderson, 1991; Cobb, 1993; Hueftle et al., 1983; Malcom, 1976). Research is clear in showing that race and gender discrimination continue to exist in the U.S. education system—both in general and in science (Pearson, 1985; Thomas, 1997; Felice, 1981; Feagin et al., 1996; Hanson, 1996; American Association of University Women, 1992; National Science Foundation, 2000). In one of the few studies of adolescent African American women's perceptions, Olsen (1996) found that young minority women see considerable advantage for whites and for men in a number of contexts (including schools). These young women view this advantage as an obstacle to their own goals. Others have found that young African American women report a sense of feeling unwelcome in the science classroom (Cobb, 1993; Vining Brown, 1994; Clewell and Anderson, 1991; Malcom, 1976; Hueftle et al., 1983). The limited research on the perceptions of women in science education and occupations also suggests that women, in general, are aware of gender barriers in the science domain (Davis, 1999; MIT, 1999). The even more limited literature on African American women in science suggests that they are aware of both race and gender barriers but perceive race as a larger barrier than gender (Rayman and Brett, 1993).

There is a growing (although still limited) body of research that suggests that in spite of the barriers that the science system sets up for women, minorities, and minority women, one cannot assume that members of these groups will be equally disinterested in science. Some research shows that African American youth, in general, hold more positive attitudes about science than any other subgroup (Hueftle et al., 1983). The research also shows that African American girls, in particular, are very positive about science (Creswell and Exezidis, 1982), sometimes even more so than their white counterparts (Mau et al., 1995). My earlier research (Hanson, 1996) looked at over time experiences in science education among those who had shown promise in science. I found young African American women to be less likely than their white counterparts to maintain high science achievement. These young women were as likely (if not slightly more likely), however, to continue being interested in science and to take science courses from the 10th grade through the post–high-school years. Interestingly, African American women are more likely than are other women of color to hope to become mathematicians (another male domain related to science) (Duran, 1987; Kenschaft, 1991).

Recent data from the National Science Foundation suggests that experiences in science from postsecondary school through occupations are distinctly different for women from different race/ethnic groups (National Science Foundation, 2000). The data suggest that increasingly, African American females are present in science. For example, African American women (but not white women) earned more than half of the Bachelor's degrees in science and engineering awarded to their race/ethnic group in 1997. During the same year, African Americans were the only race group where women earned more than one-half of the Master's degrees in science and engineering. African American women earned 46% of the Ph.D.'s awarded to African Americans in science and engineering in 1997, while white women earned 38% of the degrees awarded to whites. Finally, data from the National Science Foundation suggest that African American women make up a much larger portion of African American scientists (36%) than is the case for white women (22%). It should be noted that many of these percentages were not presented in the body of the National Science Foundation report but rather calculated from figures in the Appendix. When these figures are examined with a focus on percentages of women within race groups that are pursuing science degrees and occupations, the results often suggest a distinctly higher representation among African American women relative to white women. It is important to keep in mind, however, that although the percentages are sometimes higher for African women than for white women, the raw numbers are often very small. For example, although (within race/ethnic groups) African American women earned a larger percentage of science and engineering Ph.D.'s than did white women in 1997 (shown above), the actual number of African American women and white women earning these degrees was 280 and 5,180, respectively (National Science Foundation, 2000).

Thus, one cannot assume that African American women will be more discouraged in science than will be white women. That is not to say that these young women avoid racism and sexism in their pursuit of science. Existing studies of African American women in science document considerable barriers that these women face (Kenschaft, 1991; Malcom, 1976; Ovelton Sammons, 1990; Vining Brown, 1994). The potential for continued interest in and engagement with science in the white male science domain is what is examined here.

It is important to note that most research on minority women's science experiences has *not* found them to show higher science *achievement* than other young women (as measured by grades and standardized exam scores, especially in the high school years) (Hanson, 1996; Catsambis, 1995; Clewell and Anderson, 1991). These differences mirror race trends in the broader achievement literature. Testing biases that favor middle-class white students are most likely part of the explanation for differences in standardized exam scores (Lomax et al., 1995). Ogbu (1978, 1991) has argued that African American students' achievement may be lower than that of whites because they see fewer returns and a biased system. Thus, they reduce their efforts. Mickelson (1990) adds to this by suggesting these young people believe in education in general but not for themselves, thus predicting poorer performance. Ainsworth-Darnell and Downey (1998) have shown that it may not be so much frustration and lack of hope that work against African American students' school achievement and for white middle-class students' achievement, but the lower resources for success (including poorer schools and higher rates of poverty) that work to keep these race differences in place. Finally, race differences in achievement are most likely maintained by stratification within schools involving tracking and differential learning opportunities (Cooper, 1996; Oakes and Lipton, 1992). All of these factors, along with evidence of racism in the school and classroom (Persell, 1977; Anderson, 1988; Feagin et al., 1996) are likely to affect race differences in achievement and are relevant to my research on young African American women in science although none carefully considers gender and subareas of the curriculum. Nevertheless, the research on race differences in women's science experiences reviewed above is largely consistent with these general trends showing positive attitudes but lower achievement. However, it also shows high rates of access (course-taking) in the sciences among young African American women. Issues involving the role of schools and teachers in affecting young African American women in the school and science domain are explored further in a subsequent chapter.

What about gender patterns in the science experiences of white and African American youth? The limited research that is available suggests different gender effects. Although some research shows girls of all races doing better than (or on a par with) boys on science grades and achievement early on (Catsambis, 1995; Jacobowitz, 1983), the trend tends to reverse itself in the white but not the African American communities as the young people enter high school. For example, in the related area of math, although overall gender contrasts show boys in more advanced high school math courses than girls, one study showed that among African American youth, the opposite pattern prevails (Matthews, 1984). Interestingly, Maple and Stage (1991) found that minority females and white males were similar in their attitudes about math (and that these attitudes were important predictors of math/science majors). African American girls have been found to have more positive attitudes about math (Hart and Stanic, 1989), to be in more advanced math courses (a requirement for science degrees) (Marrett, 1981; Matthews, 1980), to get better science grades (Clewell and Anderson, 1991), and to participate more in science than their male counterparts (Marrett, 1982).

Findings From NELS

What do the analyses of NELS data collected by the National Center for Education Statistics on a nationally representative sample of high school youth tell us about the experiences of African American women in science?[1] Researchers and federal agencies that monitor science achievement in the United States have suggested that three areas of science are important dimensions of students' science experience. These include access (e.g., course-taking), achievement (e.g., grades and standardized test scores), and attitudes (National Science Foundation, 2000; Oakes, 1990; Hanson, 1996). As noted earlier, my special emphasis is on science education in the high school system, although I do consider later science experiences.

The most important goal here is to describe these science experiences for young African American women and see how they vary from those of the majority group—white women. My main focus is on differences in science experience between groups of women. Table A.3.1 in the Appendix shows figures for the longitudinal NELS sample in their 8th, 10th, and 12th grade years of high school.[2] It also provides information for the post–high-school years with regard to postsecondary education experiences and job experiences. The discussion here focuses more on trends than on specific numbers.

Science Access

In the area of science course-taking, there are a number of variables on which there are no race differences and a number on which there is an African American advantage. There are fewer comparisons showing a white advantage than an African American advantage. Results suggest that in the 8th grade, young African American women are much more likely to be in advanced, enriched, or accelerated science courses than is the case for young white women (40% vs. 24%). In 10th grade, there are no differences in course-work in chemistry. Similarly, in 12th grade there are no race differences in the young women's enrollment in science classes over the past two years, although young white women are more likely to be taking a science class at the time of the survey. When those attending college report (two years out of high school) on course-taking and majors, there are no differences in chemistry course-taking. However, young white women are more likely to have taken a physics course. But, it is young African American women who are more likely (almost twice as likely) to report a science major (at their first college/

university attended, 25% vs. 14%). When the young women are eight years out of high school, there are no race differences in their report of whether or not their first or second degree was in science. The African American women, however, are considerably more likely to report that they would like a degree in science by age 30 (26% vs. 15%).

Science Achievement

A white advantage on science achievement was expected. When occupational achievement (not school achievement as measured by grades or standardized test scores) is measured, however, this white advantage disappears. Starting in 8th grade, the young white women are more likely to get higher grades in science (in 10th grade, as well) and to score higher on standardized science exams (in all three high school years). The race difference in science exam scores is consistent across the three measurement periods. In results not presented in the table, I found that, although only a small percentage of girls from either race group received an award for a science or math fair, the young African American women are significantly more likely (twice as likely, 4% vs. 2%) to have won this award. Eight years out of high school (2000) it is young African American women who are more likely to report a current/most recent job that was in science (22% vs. 17%). All other job reports (measured two years out of high school) show no race difference.

Science Attitudes

Means for variables measuring science attitudes reveal a distinctly positive attitude toward science on the part of African American women in the early years of high school. These young women, when asked in the 8th grade (1988), are more likely than young white women to look forward to science class. In the same year, they are more likely than are young white women to feel that science will be useful in their future. There are no differences between the groups of young women on being afraid to ask questions in 8th grade science classes. Two years later (1990), in 10th grade, young African American women are considerably more likely than young white women to say that they often work hard in science class (64% vs. 55%). However, by the last year of high school (1992) these patterns shift a bit. In 12th grade, young white women claim more interest in science (although this difference is quite small, 49% vs. 42%). In the same year, there are also no race differences in respondent reports on whether they do well in science. However, young African American women are more likely to think they will need science for jobs after high school. And in 2000, when the young women have been out of high school for eight years, almost one-third (31%) of the young African American women report that the occupation that they plan to have at age 30 will be in science. Less than one-quarter of young white women report these plans (24%).

In another report, NELS was used to examine gender differences in science experiences among African American youth in the high school years (Hanson and Palmer-Johnson, 2000).[3] Interestingly, in this study of high school science experiences, we found fewer gender differences (among African American youth) than race differences (among young African American and white women). In 8th and 10th grade, young African American men score higher on standardized science exams than their female counterparts, but young African American women get higher science grades in the 8th grade. Most of these gender effects disappear when variation in other sociodemographic characteristics are taken into account. My research (Hanson, 1996), and that of others has been consistent in showing that, in general, gender differences favor boys on standardized science exams. However, gender differences are much smaller (and sometimes favor girls) on science grades. With regard to course-taking, I found that young African American men are more likely than are young African American women to be in advanced science classes and computer education classes in the 8th grade. Young African American women are more likely than are their male peers to be in earth science classes in that same year. In the 10th grade, they are still more likely to be in these courses as well as courses in general science and computer education. There are no gender differences in the course-taking behavior of these African American youth in the 12th grade. Interestingly, no differences in any of the science attitudes held by these young men and women were found. Thus, there is no clear male advantage in the NELS data for young African American students; sometimes the young women do better than the young men.

Findings from the Web Survey

Quantitative

In Table A.3.2, information on a number of science experiences that were measured as quantitative variables in the Web survey are provided.[4] Some measures are intentionally similar to those in NELS and others are unique. Keep in mind that this sample is a more current sample (surveyed in 2003) and they were included in the sample as a part of this research on minority women in science. In spite of these differences, the results from the Web survey are consistent with the NELS data. There are significant differences by race on most of the science outcomes. On some outcomes, the young white women score higher. As in NELS, they report receiving higher grades in science. Like the "interest" question in NELS, the young white women here are more likely to say they like science. Additionally, there is a significant difference favoring the young white women in the Web survey on the "good in science" question although there was none in NELS on the "does well in science" question. Consistent with the NELS data, young African American women in the Web survey are more oriented toward science occupations (relative to the young white women). They are more likely to expect a science occupation at age 30 and they are more likely to hope for a science occupation at age 30. Thus, the Web survey suggests that the young African American women's hopes, as well as plans (or expectations), are more oriented toward science occupations than are those of their white counterparts. This finding should be considered in the context of another finding, however. When asked whether they felt welcome in science, 80% of the young white women but only 66% of the young African American women agreed. An examination of the qualitative data will lend more insight into this aspect of science experience.

Two other questions asked of the respondents speak directly to research questions of interest and are unique to our Web survey. One inquires whether the respondent's interest in science is influenced by others. This addresses the issue (discussed earlier) of whether these young women see the influence of science structures (i.e., counselors and teachers) and other structures (family, community, peer) in affecting their interest in science. The other question asks the respondent whether race or gender is more of a barrier in science. These results are included in Table A.3.2.

The first thing to note is that young African American women are significantly less likely than are young white women (39% vs. 51%) to say that their science interests are influenced by others. The argument that I make in this research is that African American families and communities are a major factor in providing support (and thus an explanation) for young African American women's science interests. The findings suggest, however, that only 39% of African American women see these influences. Like the low-income women in Gaskell's (1985) research, these young women tend to view their actions and behaviors as their own choices and dispositions. Ironically, research has shown that African American families encourage independence in their daughters and thus the attitude that we found to be common may be a part of this independence encouraged by families. It may constitute part of the answer to these young women's success in a domain where (according to many respondents) they do not feel welcome.

Findings in Table A.3.2 also reveal that the young African American women in the Web survey were more likely than are their white counterparts to see race as a more formidable barrier than gender in the science domain (48% vs. 35%). Although both race and gender may work against them in the classroom, this research is consistent with other research in showing that these young women view race as the more critical factor in this process (Rayman and Brett, 1993). Hence, they often report less discrimination and a greater sense of feeling welcome at historically black colleges and universities (HBCUs) (Rayman and Brett, 1993).

The research findings produced here support those of others who have found that young African American women often feel unwelcome in science (Vining Brown, 1994; Rayman and Brett, 1993; Seymour and Hewitt, 1991; Malcom, 1976). This feeling coexists with their high expectations and hopes (relative to white women) for an occupation in science. Although both race and gender may work against them in the classroom, my research confirms that these young women view race as the more critical.

Another way to gain insight into the perceptions of science held by young African American women is to examine their responses to a vignette involving a young girl in the science classroom. By giving the vignettes to young African American and white women, and by varying the target and content of the vignette on race and gender statuses and issues, an experiment was created within a survey. As mentioned earlier, there are many advantages to this vignette approach, one of which is that the young women can respond to a concrete story and not report on their own experiences. In addition, by varying the stimuli on race and gender content one can look at the relative importance of these factors. Here, I look at two quantitative measures asked of the respondents following the vignette. These results are presented in Table A.3.3.

Information in Table A.3.3 comes from analysis of variance models referred to as multiple classification analysis (MCA). The analysis tells the extent to which groups vary on the mean of the science outcome variable and whether this variation is statistically significant. Overall, with one exception, young African American women have higher scores

than do young white women. These higher scores reflect greater reporting of something like this happening to them (i.e., what happened in
the vignette), or to other women like those in the vignette. Even when
the young woman in the vignette is white, the African American women
score as high (or higher) than do the young white women. More than
one-half of the young African American women shown a vignette with
an African American woman said that others like the woman in the
vignette do not feel welcome in science. These results are surprising
and unique to this study. They present a first-time measurement of this
phenomenon. *More than one-half* of young African American women
in a nationally representative sample say that women like those in the
vignette (African American) *do not feel welcome in science.*

When young white women were given the same vignette (with the
African American woman talking about race discrimination in the science classroom) their scores were lower than those for African American women given the same vignette (e.g., 36% said that others like
those in the vignette don't feel welcome in science). Nevertheless, this
is a rather significant number of young white women who do see race
affecting outcomes in the science classroom. Thus, young white women
see this race discrimination in the classroom as well and *more than
one-third* say that young women like the one in the vignette do not feel
welcome in science.

There were high scores on both science outcomes for young African American women shown a vignette with an African American
woman talking about race in the science classroom, suggesting that
these girls do *see* and *experience* problems. However, an unexpected
finding is that the highest scores in the table are those for African
American women who viewed a vignette showing an African American
woman but was neutral in its race/gender content. She just suggests she
feels uncomfortable in the science classroom. Fifty-three percent of the
African American women given this vignette said that this had happened to them. This figure is considerably larger than the 31% who said
this had happened to them after seeing the vignette with the African
American girl talking about race and the 16% who said this had happened to them after seeing the vignette with the African American girl
talking about gender. A similar trend involving the neutral vignette is
observed in the white sample. Here, again, the highest support for most
of the responses came with the vignette where the young white women
saw a young African American woman in a neutral vignette.

This finding produced one of the important lessons learned from
the application of the vignette method—the value of the more neutral

(or control) vignette. This control vignette contained a content that did not explicitly raise the issue of race or gender as a problem. Young African American women given this vignette were expected to report fewer barriers than those given a vignette where race was presented as an issue. Unexpectedly, the African American sample given the control vignette with the African American target reported the highest feeling of not being welcome (regarding self and others like those in the vignette) of any group in the study. Young white women in the sample also reported the highest scores on the "unwelcome" variable when they were given the control vignette involving an African American target. These scores were even higher than when the white respondent was shown a young white woman (like themselves), who talked about gender issues in the science classroom. Thus, respondents may be more likely to see barriers when they conclude them on their own, as opposed to situations where race or gender are explicitly brought up as issues. Originally, I speculated that including race or gender in the content of the vignette might lead young women to see these statuses as problematic. Results suggest that they lead the respondent in the opposite direction.

With regard to perceptions of gender, the young African American women given the vignette with a young African American woman talking about gender tended to report more problems for girls like this than did the young white women given the vignette with an African American woman or a white woman talking about gender. This supports the notion that young African American women see not only race but also gender as a larger problem in science than do young white women. They see this as a problem for both African American and white women. However, the young African American women who received the vignette with the young African American talking about race scored higher than those receiving a similar vignette focusing on gender. The young white women gave very similar responses to the race-content and gender-content vignettes. Thus, although young African American and white women see both race and gender barriers in science, African American (but not white) women see race as a larger barrier.

It was expected that young African American women would be more likely to see barriers and report feelings of being unwelcome for other women than for themselves. Indeed, the percentage of African American women who felt others like those in the vignette do not feel welcome in science is larger than the percentage reporting that something like what happened in the vignette had happened to them. Note that the young white women given a vignette with a young white woman also see more problems for other white women than for themselves in the science classroom.

The group that seems to feel the least unwelcome and see the fewest problems (for self or others like those in the vignette) tends to be the white women who were given a vignette with a white woman and a gender content.

As expected, the race of the target affects African American girl's reports. Young African American women report higher scores on questions asking whether this had happened to them when the target in the vignette is African American. Interestingly, young white women were not as affected by the race of the target and saw, for example, similar problems for young African American and white women in most (but not all) of their responses to the vignettes focusing on gender.

The responses to the vignettes in the Web survey provide valuable insight into the complex ways in which race and gender affect young women's perceptions of the climate for young women in the science classroom. More than one-half of the African American women in the sample suggested that African American women (like the one in the vignette) do not feel welcome in science. They felt that race was a larger barrier then gender in creating problems for young women. Next, let us turn to the qualitative portion of the survey. Here, we will learn more about these perceptions by looking at the young African American women's thoughts, in their own words.

Qualitative

In this section, I will report some of the young African American women's thoughts on their experiences in science as reported in the Web survey. First, I focus on the survey questions that followed the vignette: What is your first response? Have you ever felt this way? Are others like LaToya interested in science? Do others, like LaToya, feel unwelcome in science? The focus is especially on the African American sample who responded to the vignette where LaToya talks about her discomfort with being black in the white science classroom. Finally, I consider responses to an open-ended question not associated with the vignette. Here, I look at all African American women in the sample. The question asks why the respondent likes (or does not like) science.

Responses to the Vignette
The first question asked after the vignette was, "What is your first response to LaToya's experience?" There are a number of different

types of responses to this question. First, there are a good number of respondents whose first response was one of *empathy* … this had happened to them as well. These responses often reflect an awareness of racism in the science class. Another common response is of *sympathy* or *acknowledgment*. They feel bad for LaToya. A majority of the responses reflect empathy or sympathy for LaToya. Sometimes the first response is a factual statement about the existence of racism in the classroom. A third type of response (one of the most common) is to give *encouragement and advice*. Some advice is very specific and pro-active. Many of these young women are aware of the contributions of African Americans to science. Other advice involves recommendations to just hang in there and persevere; some suggest that LaToya will eventually get a teacher who cares. Still other advice suggests that LaToya should disregard what others think and just follow her dream. There are a small number of respondents who seem to lay the blame on LaToya not applying herself or trying hard enough. They suggest that she get beyond the color issue. Finally, there are some (again, a small number) who are very upbeat and their first response is simply that they *recognize LaToya's interest in science and admire it*. Some of these responses are shared in Table 3.1. Note that some young girls give responses that include comments falling in more than one of these categories.

One may be struck by the large majority of respondents who suggested in their "first response" to the vignette that they knew (from first-hand or second-hand experience) what LaToya was going through.

TABLE 3.1 Response to Vignettes: "What Is Your First Response to LaToya's Experience?"

Empathy

1. "Many black women feel the same way."
2. "Much like an experience I had … [at a high school in] the suburbs … the teachers did call on you and when you had the right answer the students would turn and look at you in shock."
3. "I understand completely, I was the only black in my physics class and … I just never fit in there … it did affect my grade."
4. "I can relate."
5. "I experienced the same thing in high school."
6. "This is something that I can relate to considering I went to a predominantly white university. I was discriminated against in different ways."
7. "I thought about how I felt when I was in high school biology. I felt that I could never do as good as I hoped because the teacher was talking on a Ph.D. level and I was lost."

(continued)

TABLE 3.1 *Continued*

8. "Well I can relate to LaToya. I went to a high school that was predominantly white and can make an African American feel out of place. It can make you feel as though this particular subject is not for us."

9. "I felt the same way, not in high school, but at my first undergraduate institution. I was practically invisible to several of my professors. However, moving on through life, I have discovered that it's not quite that bad everywhere. It's still difficult to be African American, female, and pursuing a career in science, but the resolve of doing what God has placed me here to do is well worth the adversity."

10. "I experienced that same feeling except that I do get called upon in class because I go to a very small school. Sometimes I do feel out of place because I am a student of color in an all white school and I do see only white scientists in my text books and in the films we watch but that only makes me want to be the best student I can be so that people won't think that only white people can be scientists."

11. "I felt the same when in my science classes in high school. In most of my classes I was the only African American, and that made me feel uncomfortable."

Sympathy/Acknowledgment

1. "Many black women feel the same way."
2. "She's right most black students don't feel they can accomplish much."
3. "It's true."
4. "My first response is that LaToya's experience is horrible and she should do something about that she shouldn't be discriminated against just because she is black she should do something about that A.S.A.P."
5. "I feel bad for her."
6. "The world can be prejudiced."
7. "She thinks she is not wanted."
8. "I agree with everything she had to say."
9. "I don't know maybe she has a very good reason to feel as she does but I don't get those feelings at my school."
10. "I can understand her point of view and could see that happening to her. There aren't too many African Americans in the field of science. So, naturally, most Caucasian teachers may feel LaToya may not excel in science as well and therefore do not look for her great participation."
11. "It's too bad you're left out."
12. "Typical!!"
13. "This is very common to feel like LaToya in everything blacks are involved in from sports to music, acting, fashion, and educational. We are made to feel as if we are not good enough. However we end up becoming the best at these things."
14. "I think LaToya has a point in saying that she doesn't feel as though African American's are portrayed as major participants in the science of America, but I think this is true in just about every subject. Most textbooks, for whatever reason, tend to put white people in their books, maybe because more people will relate to those pictures. If African Americans let who they see in books affect what they decide to become then we will rarely amount to anything large, and therefore will not be worthy as pictures in the text book. LaToya should follow her dreams regardless of something as petty as pictures in a textbook."

TABLE 3.1 *Continued*

Encouragement/Advice

1. "LaToya should not feel intimidated by what she sees because if she loves science she should pursue it and have faith in God. She loves science and if she's not called on I just would like to encourage her to listen and take in all of the answers that are given, and God will give her a chance to become a great scientist."
2. "You can be a black science person and when you make it thank God because you are black."
3. "There is no reason for her to feel that way considering this is the year 2003. African Americans are doing everything. I feel that if she knows science pretty well then she shouldn't stop learning because of what she thinks others expect of her."
4. "Don't let anyone tell you where you don't belong. Follow your dreams. Keep going and you will find teachers who care about you."
5. "I think perhaps LaToya should learn more about her history. Perhaps then she would find the first scientists, mathematicians, astrologists, architects, and the list goes on, were black, and she would be able to relate to the subject at hand. Never mind what the scientists in books and her classmates look like."
6. "Not applying herself if she feels left out."
7. "Maybe she should transfer."
8. "She needs to look over the 'color' issue and put her future plans into action."
9. "If she loves the field, then she should do as much as she can to make her persistence known in her class and not wait to be called on. Many pioneers in science were and are black, and books and classroom teachings should reflect that."
10. "I feel she should take science and not let that teacher ruin something she loves; I would address the issue with the principal of the school."
11. "If she loves science, she should stick with science."
12. "Just worry about LaToya. No one else is going to pay your bills when the time comes so don't let that feeling hinder you in any way. You don't always have to 'belong'. Just get your education and use it to your advantage."
13. "Initial response was 'girl don't let that discourage you from becoming a scientist'. We need more young ladies like yourself."
14. "I encourage people like LaToya. They are the ones that make us proud."
15. "I think LaToya needs to talk with her teacher to try and find out what the problem is. Why is she not getting called on?"
16. "Get more involved."
17. "LaToya does belong in science. She has to be ready to answer questions."
18. "I think she shouldn't worry about what color people are sitting next to her, just be happy she's doing something she loves to do. Maybe she should start raising her hand to answer questions, so the teacher can know she wants to share her thoughts and answers with the class."

See LaToya's Interest in Science

1. "She loves science."
2. "That she loves science."
3. "Bright student who loves science."
4. "I think its great for what she is doing."
5. "I admire LaToya for her interest in science. I think we need more African American scientists."

They talked about feeling "invisible" in the classroom. The responses also reflected some hope. Many of the young women, sometimes even those who reported similar negative experiences, gave advice and encouragement revealing a source of strength and pride in African American accomplishments and in African Americans who are scientists. In fact, one girl reports that blacks are "made to feel as if we are not good enough" but then goes on to say "however we end up becoming the best." The advice often suggests rising above the color issue, not focusing on racism (although they were very aware of it), and not worrying about "belonging." Recall that LaToya ends her story with "Sometimes I don't feel like I belong in science." For some women the source of strength that they tell LaToya to rely on comes from religion. Sometimes, the strength seems to come from a desire to disprove the notion that "only white people can be scientists." It is also clear that these women have faith in education. They give LaToya considerable encouragement to continue in science education in spite of the costs. No one advised her to give up. Only a very small minority of women did not provide a response or gave responses such as "I don't care." Another small minority said that they had not experienced this feeling of not belonging in science and that African American and white teachers treated them the same. One young woman had no response because "my school was predominantly minority so we didn't have any racial problems." Only one of these respondents reacted to LaToya's gender in her answer. In fact, for most, it was race and the difficulties African Americans face in schools and science classes that they focused on. As the results from other parts of the survey confirm, race is a larger issue for these young women than is gender. One wishes (and hopes) that young African American women in today's science classrooms could read these pieces of advice. Part of the answer to the research question involving young African American women's interest and engagement in science (in spite of racism and sexism) can be seen in the agency that many of these young women express.

Two other questions following the vignette inquire *whether anything like this had ever happened to them* and *whether other young African American women are interested in science*. Again, I focus on the group of African American girls who responded to the vignette with the young African American woman talking about race in the science classroom in my discussion of these answers.

Thirty-one percent of African American women given the vignette with LaToya talking about race reported that *something like this had happened to them*. When asked to elaborate, the young women talked about things that had happened to them in the classroom. Many of these responses are discussed in Chapter 4, on teachers and schools, but the more general response to this question often has to do with how LaToya's experience is the norm, how these young women do not feel they belong in science, and how science appears to be a "white" enterprise. As to the issue of LaToya's experience being the norm, one young woman reports: "There are too many examples to name. It happened all the time in my classes in high school, college, and is continuing to happen even in my managerial position at our Gas and Electric Company."

Another type of response focuses clearly on the bias in science. One young woman claims, "In my science classes I only learn about white scientists."

The feeling of not belonging is clear in several girls' remarks:

> "It seemed to me that as an African American I should not have been in this particular class."
> "You just feel weird."
> "I, too, felt like I did not belong."

Seventy-two percent of the sample of young African American women given the vignette with LaToya talking about race said that *other students, like LaToya are interested in science.* When asked to elaborate on this the girls often talked about an interest in science but barriers that made this interest difficult to pursue. Sometimes they just gave an upbeat, positive answer to this question. Several typical comments reflecting interest but barriers are:

> "Yes, science is in the heart of black people we just need to given the help … the right tools."
> "Yes, I think science is interesting … if it is understood it can be fun but a lot of black high schools are not teaching science like it should be taught."
> "Yes, I have several friends that went to school for computer science they have degrees but were not able to find jobs in that field."
> "Yes, African American students are interested in the full spectrum of subjects. Science is no exception. The difference is that they may not be encouraged as much to pursue science careers."
> "Yes, I think there are a lot of people out there like LaToya. But out of African Americans as a whole there is a stereotype that most African Americans don't like science but some do."

Responses that were upbeat about the interest of others like LaToya include, for example:

"Yes I've known many students who went on to college and careers in the science field."

"Why??? Why not??? Is there a reason black students shouldn't like science?"

"Yes, African Americans are breaking into all major areas of different fields, science included."

"Yes. Because I don't think it matters what color you are if you like science, you like science, if you don't, you don't."

"Yes, because I know a lot of African Americans who are interested in science."

"Yes, because they want to learn just as a white kid would."

"Yes, because I think anyone can be interested in science."

"Yes, I think a lot of African Americans like to learn about how things formed and why things are the way they are."

Reasons for Liking (or Not Liking) Science

Finally, a general question was asked of the young women about their reasons for liking (or not liking) science. The young African American women in the sample were very thoughtful and detailed in their responses. Some of them were quite eloquent in their reply to this question. One young woman reported, "Science is like opening a present from your favorite aunt. You just can't wait to open it because you know that there is something wonderful and unique inside."

Another aspect of attitudes toward science that became clear in the examination of the young women's response to this question on liking science was the diversity of reasons given. No answers were exactly alike. Even though many answers reflected similar themes (e.g., science is too hard), the way that the attitude was expressed and the combination of reasons given brought us to the realization that these young women have had diverse experiences and exposures to science. Two young women might, for example, both report having bad teachers in science. However, one (and not the other) might report that her mother did experiments with her at home and made science fun. Thus, although the major themes for liking and not liking science are presented in the sections below, this complexity and diversity in responses should be kept in mind. Additionally, the complexity of answers provides insights into ways of teaching science that create a love for science and ways that make it boring. For example, two young women in the sample talk about the same subject, physics, in very different ways. One hates it because of the fact that she has to sit and memorize formulae: "I don't like science because it is based so much on memorizing. In biology, I spent a lot of time just memorizing the make-up of the cell or DNA or things like that. Physics spent a lot of time memorizing formulas for speed, velocity, acceleration, etc."

Another girl says that her favorite subject was physics because of how the teacher taught it. "I like science.... I think my favorite thing this year was learning about force and motion because we got to make roller coasters and it was a plausible way of going about studying that particular section."

In each of these cases, the young woman's overall evaluation of science was affected by this experience in the physics classroom. The implications of this for who teaches science and how they teach it are obvious.

When considered as a yes/no question, only one-third (34%) of the young women reported that they liked science. This question was followed with an open-ended item that asked, "Why do you like, or not like, science?" The answers to this question suggest that many of the young women actually have mixed feelings about science. One of the lessons learned from this finding is that young people's feelings toward science are not reflected in a single question on how much they like science. Likes and dislikes are complex things that seldom follow simple dichotomies. Researchers interested in measuring attitudes toward science need to allow young people to reflect this complexity in their responses. If researchers use quantitative measures it is important that they include questions about what they like about science and what they dislike about science without simply asking whether they like or dislike science. The young women in this sample were often very specific about what they like about science and what they don't like. For example, one young woman reported: "I like science because it gives a lot of information on the human anatomy, which will be very helpful in my nursing career. I don't like science because of all of the hard names of diseases and medicines that I would have to remember."

The young women are also very clear on the fact that they like some areas of science but not others. Again, simple questions that ask about science overall neglect the fact that there are many areas of science that involve different topics and skills. For example, one young women in the sample reports, "It depends on what area of science. For an example, physics teaches more about conversions, energy, kinetic and potential motions, etc. While on the other hand you have courses such as anatomy teaching on the body (heart: atrium, ventrium, arteries, etc.). I have taken nearly all areas of science throughout high school ... and have found that I am more interested in anatomy and environmental/

earth sciences more so than physics or chemistry because these two particular courses pertain to math."

Another young woman reports: "I really enjoyed and did well in biology, and really did not enjoy nor understand chemistry. Chemistry was very abstract to me and did not give me an 'aha' experience that other subjects gave me."

The discussion below reports on some of the reasons that the young women in the sample like or dislike science but it is important to keep in mind that some of these reasons were combined in single answers. Examples of these answers will also be provided.

There were a number of themes that were reflected in the reasons given for *liking science*. Although it was not a common theme, some young women gave answers that involved the spiritual. Several examples of these responses are: "I find the intricate details of cellular processes, including physiological and biochemical pathways fascinating. They illustrate for me, in a natural sense, a taste of the awesome nature of God." "I like science because it is fascinating. I love learning about how we are wonderfully made as humans. I think it proves that there is a God."

Some young women like science but make it clear they would not like a career in science. For example, one young girl answered the question: "I feel science is very interesting. But, I would never like to have science as a career."

Other young women said that it wasn't a question of liking or liking science, it was a necessity: "It's not that I like or dislike science, it is something that I will need for the career that I chose."

Other young women provide answers about a career in science that are a bit more perplexing: "I really don't [like science] but it is what I want to do."

It was interesting to read the responses from young women who just enjoyed the scientific process. One young woman replied, "I like science because it is rational. Events occur in nature or a laboratory and they can be explained with data. I like the whole process of making a hypothesis, then finding evidence to support or repudiate your claim."

The most common themes for liking science involved liking science because: 1) it answered questions relevant to life and nature and the everyday world; 2) it was just fun, largely because the person was a curious person and many possibilities were explored through science and the imagination; 3) it involved doing experiments and hands-on activities. Other themes that were less frequently sited were: 4) science is a challenge; 5) an enjoyment in taking things apart and dissecting;

6) they just like it; 7) it's cool; and 8) they are good at it. Some examples of the young girl's answers reflecting each of these themes are provided in Table 3.2.

I also examined the reasons the young girls gave for *not liking science*. One of the most common themes here had to do with 1) teachers that did not make science interesting. Other common themes for not

---

TABLE 3.2 Reasons for Liking Science

Science Answers Questions (and Provides Solutions) Relevant to Life and Nature and the Everyday World

1. "I like science because it is interesting and exciting learning about my surroundings."
2. "I think the new studies in medicine are great, the cures for diseases, etc."
3. "Very much, science helps me understand the structure of nature and the life of man and animals. I like explorations of the universe. There is nothing I don't like about science."
4. "Very much because it [helps one] understand the world we live in and understand the substance that makes the world work."
5. "Very much. I like science because it is life. What would the world be like without it? Science puts everything into perspective."
6. "I love biology for the simple fact that it is the study of life. I enjoy studying the human body. I do not like chemistry, but I know that this is something I will have to overcome."
7. "It's very interesting. In my anatomy class we're learning about the body and they break it down section by section. I'm starting to use medical terms now in my everyday life when I didn't do that before. It also puts me more in tune with myself."
8. "Its an important part of everyday life."
9. "I like science very much because of the fact that it answers commonly asked questions and enlightens my mind."
10. "I like science because it gives you insight into how certain things in life work, live, and reproduce."
11. "I like science because there are not limits with what you can do. It makes life easier and it's good to understand how it is making life less complicated."
12. "I love the biological sciences, learning about the form and function of human bodies, as well as animal bodies. I dislike chemistry and physics because of the abstractness involved. I am far more concrete."
13. "I just like the subject because it can help explain things."
14. "I like science because it allows you to do research to figure out something and lets you mix different chemicals or look at organisms under a microscope."
15. "I like science because it's very educational and [I] will need it in the future."
16. "Science is okay as long as it applies to my nursing classes but other than that I hate it."
17. "It's a living study. It is something you experience every day of your life. It's very practical and tangible."

(continued)

TABLE 3.2 *Continued*

Science Is Fun and Interesting, Especially for Those with Curiosity and an Imagination

1. "I like to learn about how things function."
2. "Exploring theories and using formulas to prove them."
3. "I enjoyed the teacher, she made class fun."
4. "Very much, it's fun."
5. "It is interesting, fun to learn."
6. "I like science because when you study it you kinda warp into the world of creatures, plants, and other awesome stuff."
7. "Very much. It deals with amazing things that can't be explained … the body … cells … reproduction … force … atoms … things like that."
8. "Very much. I always enjoyed what made things happen. Science provides the answers."
9. "I enjoy science because it makes sense. The process for solving a problem is simple and with it you can solve many complex problems. Science is also very dependent on math, which is my favorite subject."
10. "The reason why I like science is because I like to learn about the different types of animals and the different types of insects."
11. "Very much, it is a subject that interests me a lot and is fun."
12. "Very much. Discovering cures or interactive drugs for various terminal illnesses."
13. "I have always loved and been interested in how things work and I have wondered why things are. I am currently really kind of obsessed with food science. I had sort of planned to be a scientist until I got to my high school chemistry class. As opposed to being all about why things work they way they do, it was all about this chart that we had to memorize and regurgitate back to the teacher. I truly hated it."
14. "Some areas of science are very hard to understand and do not catch my attention, while other areas are very interesting. Forensic science is very interesting to learn."
15. "Science is interesting when learning about why things happen, i.e., weather, seasons, exploration, forensics."
16. "Science is amazing because there is an answer for everything! Also, once you know the answer, science is predictable. Anything that is a mystery, is just that. Mysteries have solutions and a scientist is just a detective."
17. "I like science because there are not limits with what you can do. It makes life easier and it's good to understand how it is making life less complicated."
18. "Very much. I like it because the 'unknown' can be explored and examined through science."
19. "Very much, because it just interests me."
20. "I like science because there is always something to be explored."
21. "I like the atmosphere and that it makes me understand more about nature."
22. "Details on animal and insect habits. New studies in the deep sea. Sometimes I enjoy new studies in space exploration. Finally, I like forensic science and biology."
23. "Very much. I like the possibilities of science. The numerous outcomes and answers and how things could vary with the change of the slightest data input."
24. "I've always been interested in why things work the way they do and how."

TABLE 3.2 *Continued*

Science Involves Experiments and Hands-On Activities

1. "Very much because we get to do experiments in her class."
2. "Because of the different experiments."
3. "Science is okay. I like working in the lab. Working with chemicals."
4. "I like mixing things together. [But] remembering the equations can be hard."
5. "In my science class we always do hands-on things like labs. We test for lipids in blood. We are always on the move. Plus my teacher gets us enrolled in the state-wide science olypmia."
6. " I like doing some of the experiments."
7. "Very much. I like the hands-on activities and experiences that go along with science."
8. "I like understanding how things work ... anything from computer equipment to enzymes in our bodies. It is very interesting. Just sometimes the math in the equations turns me off."
9. "The experiments. Exploring different types of things."
10. "I do not like earth science. I like science when you have to experiment with things or objects to find out why something happened."
11. "I like the experiments and math work."
12. "I like science because, you get to experiment. Also because science is everything around you, so you're really experiencing science in everyday life."
13. "I like the lab portions of science, learning how things really work and how they are made."

Science Is a Challenge

1. "While in high school I was in numerous science fairs. I liked the challenge of what I was learning about."
2. " I like science because it's fun and it gives you an opportunity to go further than what you expected you would in class."
3. "Very much. It's a challenge."
4. "Very much. I really love a challenge. Science involves a lot of thought, and that is why I like it."

Science Is Just Something I Like

1. "I like science."
2. "Just like."

Science Is Something I'm Good At

1. "I always excelled in science and all my teachers were wonderful."
2. "I like science because first of all—I'm pretty good at it. I think the subject jells with the way I think. I'm not a very good memorizer. I do better when I understand why and how things are the way they are and then I'll never forget it. This is why history and literature are not for me."
3. "It's easy."

(continued)

TABLE 3.2 *Continued*

Science Involves Taking Things Apart and Dissecting

1. "I like science. I like the blood and gut part."
2. "I like science. I loved dissecting animals and mixing different chemicals."
3. "I like to dissect things and know about things on the inside and science gives you that."
4. "I love science. It's so fascinating. It teaches me different things that I never knew about. I'm in biology right now and we get to dissect animals. It's so fun."

Science Is Cool

1. "It's neat."
2. "Very much. Cause it's cool. I like learning about the body and how it works. I like that I can help other kids in my class and move around and not try so hard to get the grades."
3. "Very much. I like science because you get to do all of these really cool experiments like trying to figure out an unknown chemical and also, on the unit we just finished, we got to see how different chemicals or fibers burned and the color they turned when put into a direct flame."
4. "I like it because of all the cool labs and experiments, and how it has to do so much with life and our community."

liking science involved 2) science is too hard, complicated, rigid, or intimidating; 3) science it too vague and not relevant to everyday life; 4) science is too bloody and often involves dissecting; and 5) science is boring. Two other, less common themes were that 6) science is not interesting or that 7) science is too analytic.

Some young women do not have a reason for disliking science. They just don't like it. One young woman reported that she does not like science because it leaves out the spiritual. Her reply: "I do not like that science is rigid and narrow minded and only believes in the material or things that can be detected with the senses as real. It leaves a lot of the invisible and spiritual out. It is only concerned with what can be proved. Not the spiritual."

Again, showing the complexity of responses, the same young woman goes on to say, "I like that it can save your life."

Finally, a few women said that they did not like science, in part, because of peer pressure. One woman notes, "My peers would classify each other as nerds if you showed too much interest in science. I was very sensitive to what others thought of me and I neglected science."

When the young women talk about teachers, it often appears to be the case that they might like science but their teachers did not make it fun. Many reported that they did not like the teacher and that the teacher did not teach them much. Replies of this nature seem to go hand in hand with the ones that suggest science is too hard or not interesting. Thus the answers to the question that reflected an overall dislike for science seemed to be even more interrelated to each other and involving mixed responses than those which reflected an overall liking for science. Additionally, these mixed replies involving teachers often suggest a potential love for science but bad experiences with teachers that took away from a development of science interests. The following responses provide examples of these patterns: "Science was not all that interesting to me when I was in high school. There were not too many teachers who made it interesting. I do recall one year of biology that was so much fun in which I probably learned the most in all my years in high school"; "I really don't like it that much because my science teacher never lets us do labs. She just makes us sit there and do worksheets the whole class. Plus, she never lets us burn up anything."

It is important to note that although most of the comments about teachers were part of the reason the young women did not like science, there were some young women who brought up the teacher for the opposite reason. For example, one girl notes, "Physics is a little hard for me but I pass because my parents and that teacher don't want me getting grades below C's." This young woman, as well as others, also brings up parents when she talks about liking or doing well in science. The influences of both teachers and parents will be examined in more detail in a later chapter.

An important phenomenon that was observed in this section is that when science is just presented as memorization, it does not inspire these young women. A common theme in the reasons for not liking science is the focus on memorizing chemical tables, make-up of the cell, formulae for speed, velocity, etc. Some of the young women blame themselves for being lazy and not wanting to do all of this memorization. These responses often make suggestions as to what they do like about science and here it is often hands-on activities that are noted. One young woman, who does not like to memorize, provides the perfect solution: "I'm not a very good memorizer. I do better when I understand why and how things are the way they are and then I'll never forget it." Again, these responses provide insights into the way science is taught and how it can be made more interesting to young minority women, young women, and students in general.

One of the most common themes in the reasons for disliking science had to do with the fact that science was too hard and they were not good at it. Similarly, some women in the section above reported that they liked science simply because they were good at it. The implications of these responses for the science classroom involve the relationship between perceived abilities (and confidence) in science and one's positive attitude toward science. Young people's feelings about abilities in a subject are often influenced by teachers and peers. These "self-fulfilling prophecies" in science often revolve around race and gender, as noted in the discussion above. When teachers walk into the science classroom expecting talent from every student, it is likely that the "self-fulfilling prophecy" will be one of talent development rather than talent loss.

Examples of young women's explanations for not liking science, and that fall into the most common categories noted above, are provided in Table 3.3.

TABLE 3.3 Reasons for Not Liking Science

Science Teachers Don't Make Science Interesting

1. "I think it's because my teacher can't teach. About 4 people out of 26 are passing. I guess it would be alright if I had another teacher."
2. "Not so much. I do not like science because it is difficult to learn with the teacher I currently have and the students are also trouble."
3. "I have always loved and been interested in how things work and I have wondered why things are. I am currently really kind of obsessed with food science. I had sort of planned to be a scientist until I got to my high school chemistry class. As opposed to being all about why things work the way they do, it was all about this chart that we had to memorize and regurgitate back to the teacher. I truly hated it."
4. "I don't really like science because I don't understand any of it. Maybe it's because my teachers but I don't like it, and it's the only class I'm failing."
5. "It could have had something to do with the instructors I had ... very blah!!"
6. "The teacher that taught science didn't make it fun, and it's a fun subject."
7. "Well, the reason why I do like science is because it is interesting ... you learn more everyday. The reason why I don't like science is because it is complicated and the teacher I have at my school doesn't teach us much ... and [that is] scary because when I go to college I will just have to learn it all over again."
8. "First of all, the teachers were not enthusiastic during classes."
9. "It is not the subject itself but the teachers."

Science Is Too Bloody, Involves Dissecting

1. "Don't like blood."
2. "I like some types of science like astronomy, but I hate dissecting things, and chemistry is a little difficult. But to pass my classes I learn what I have to, and do my work."
3. "I like science until we get to the dissecting of dead baby pigs and cow eyes. I feel that is disgusting and if I wanted to do that I would have already started my pre-requisites for mortuary science."

TABLE 3.3 *Continued*

4. "Not so much. [Don't like] the opening of animals."
5. "What I don't like about science is how they have to dissect the animals and tell them about the different parts of the body."
6. "Science is O.K. I don't like dissecting animals."

Science Is Too Hard, Complicated, Rigid, Intimidating

1. "It's ok. Don't like memorizing things like the periodic table of elements, too many abbreviations."
2. "The material is very dense and there is a lot of memorization. I don't feel that I ever really grasp the concepts being taught and I just try to remember information for a test."
3. "Because it's so much you have to learn with science."
4. "Too complicated."
5. "Not so much, because it's hard."
6. "Did not have a good foundation for science and math."
7. "No, I do not like science, for one you need to give 100% effort into it and if you can't you'll be wasting your time until you build your skills to be 100% ready."
8. "It takes a whole lot of concentration understanding math. I hate math."
9. "I do not like science because I never could understand it."
10. "I am really not into science. I loved to do the different experiments, however, the content of the courses are quite difficult."
11. "Too many details and memorization."
12. "I like some science but can't comprehend some."
13. "I like it, but sometimes it gets very complex."
14. "I like some of the experiments we used to do in science class but I have a problem with understanding or remembering different principles of science."
15. "Too much to remember."
16. "I don't like science because it takes a lot of brain knowledge to find out the answer to some of the problems, and you are under a lot of pressure to remember what you have learned and put it into something that may require what you learned."
17. "Some areas of science are very hard to understand and do not catch my attention, while other areas are very interesting. Forensic science is very interesting to learn."
18. "The math involved is hard."
19. "I don't like science because it's real hard to comprehend."
20. "I did very well in some areas of science and others I didn't grasp at all."
21. "Science is something that I really have to put my mind to. It doesn't come natural to me like business. I need a little encouragement in science."
22. "Just don't. I don't understand most of the questions in science books."
23. "I don't like science because it is based so much on memorizing. In biology I spent a lot of time just memorizing the make-up of the cell or DNA or things like that. Physics spent a lot of time memorizing formulas for speed, velocity, acceleration, etc. However, in chemistry, I did enjoy doing experiments because hands-on learning is always more interesting and enticing then lecture form."
24. "Not so much. 'Cause it is too much. Especially chemical mixtures and plants. Too complicated."

(continued)

TABLE 3.3 *Continued*

25. "I don't really like science, because I am lazy and science has a lot do with remembering and the way that science supposedly makes sense isn't really that obvious to me."
26. "Science is okay. It can be very hard if you do not study or have good study techniques. Anatomy killed me. The skull!!! All the bones, not to mention the muscles of the face."
27. "Too many formulas."

Science Is Vague/Abstract, Not Relevant

1. "Biochemistry, for example, is really complicated with a lot of information that I WON'T use on an everyday basis."
2. "Not so much because it is not in my future plans to be a scientist of any kind."
3. "Found business subjects had greater correlation with real life experiences than science."
4. "I just don't like science because it is boring, I don't think it's important."
5. "I don't like science because it is so vague. It is a subject of which there is really no right or wrong answer. The only purpose of science is to make discoveries or theories about life. I much prefer subjects in which the objective is to solve a specific problem that has a specific answer, such as mathematics."
6. "I don't care for science because it is not what I need to get the job I want."
7. "I love the biological sciences, learning about the form and function of human bodies, as well as animal bodies. I dislike chemistry and physics because of the abstractness involved. I am far more concrete."
8. "Because I'm not going to be using science in my near future."
9. "I did not enjoy having to know the technical names for plants, animals, etc. if my plan was to work on humans."

Science Is Boring/Stupid

1. "Science is boring and is sometimes hard to understand."
2. "I do not like science because it is really boring."
3. "It's just boring to me."
4. "I don't like anything about science. The whole subject is boring to me."
5. "It's stupid."
6. "No, I do not like science because it is boring. We don't do anything but crazy labs and boring worksheets."
7. "Because it's boring and I don't like the teacher."
8. "Not so much. It is not that interesting. I am more of a people person."

Science Is Not Interesting

1. "Doesn't really interest me."
2. "I don't hate science. It's just not my favorite subject. I do like doing labs."
3. "The sciences were just not my strong suit. I was more interested in words and how to put them together for effective communication. I loved reading and books. The numbers, formulas, and experiments in science and math did not interest me. Moreover, I found them intimidating because I had no control over them."

TABLE 3.3 *Continued*

---

4. "I never developed a particular interest in science."
5. "Science never really interested me. It requires a lot of studying, going to the lab, and researching a lot of different things. It's not something I would like to do for the rest of my life."

Science Is Too Analytic

1. "Because I don't think analytically all the time, and in science, you need to think analytically all the time."
2. "I don't like to measure quantity of anything at all and mixing things."

Conclusions

Some have argued that since women perform less well in science than men, and minorities do less well than whites, there will be a double disadvantage for young African American women in science (Clewell and Anderson, 1991; Cobb, 1993; Vining Brown, 1994). The double jeopardy argument assumes an additive effect of the two statuses—being female and African American. In this chapter, I have painted a picture of young African American women's experiences in science using recent literature and a variety of data from multiple sources. The picture is a complex one that does not necessarily support the double jeopardy argument.

Findings from the NELS data show that in the area of science course-taking, there are often no differences between young African American and white women. When there are differences, there tends to be an African American advantage. Although data on science achievement shows a clear white advantage on most indicators, eight years out of high school the African American women are more likely to report a job in science. Similarly, findings on science attitudes show a distinctly positive attitude toward science on the part of the young African American women (sometimes more positive than among white women).

Thus, many of these young women are taking science courses and expressing interest in science. They are more likely than are young white women to hope (and plan) for a career in science. The Knowledge Network data suggest that large numbers of young African American women in the sample express a fascination with science and the answers it provides to questions of the imagination and of everyday life.

Yet a significant number of these young African American women do not feel welcome in science, see race discrimination in science education, and report that science is too hard and intimidating for them. In the next chapter, I consider the role of schools and teachers in the process whereby young African American women are encouraged or discouraged in science. The conceptual framework and review of the literature suggests that schools and teachers will be more associated with discouragement in science than encouragement in science.

4 Influences—Teachers and Schools

"They Looked at Us Like We Weren't Supposed to Be Scientists."

The anomaly that is addressed in this volume centers on the positive attitudes toward science that many young African American women express in the context of a science environment that is not welcoming. In this chapter, I look at the literature on education systems (and especially science education systems) in the United States and the way that race and gender come into play. I also present analyses from the NELS and Knowledge Networks data that provide insights into the role of schools and teachers in young African American women's science experiences.

Background: The Literature on Schools, Teachers, and Minority Women's Science Experiences

The focus on education as a means to escape poverty and achieve mobility is a major emphasis in African American communities (Hanson and Ginsburg, 1988; Anderson, 1988; Ogbu, 1991; Ainsworth-Darnell and Downey, 1998). Beginning with the Coleman report (Coleman et al., 1966), there has been considerable evidence showing the positive educational attitudes of African American youth—attitudes that often persist in the context of poor achievement. It would be tempting to cite research on young African American student's "anti-academic, anti-white" school attitudes in an effort to refute this claim (Ogbu, 1974). However, race is a complex factor in the U.S. educational system and simple conclusions do not hold. In fact, there is considerable evidence that a positive value on education exists in many minority communities (including African American). Most parents of minority students have a high value on education, high aspirations for their children, and a desire for involvement in their children's academic development (Cummins, 1993). Student's values mirror these (Ainsworth-Darnell and Downey, 1998). However, school systems in the United States have race and social-class biases that limit the achievement of many young people in minority statuses. The past decades have seen improvement, but many biases (some subtle and some not so subtle) still exist (Banks, 1991; Persell, 1977; Kao and Thompson, 2003; Davis, 2004; Feagin et al., 1996). Case studies of minority college students in the 1990s revealed similar observations as those done decades earlier (Davis, 2004).

The myth that schools are an equalizing force in a country where every child has equal opportunity for success hides the reality of inequality in educational access, resources, and opportunities. Negative stereotypes (Steele, 1997), an alien physical and social culture (Anderson, 1988), and biased teachers, curriculum (Kao and Thompson, 2004; Anderson, 1988; Banks, 1991; Persell, 1977), and testing (Haney, 1993; Persell, 1977) are factors creating inequality by race (as well as class and gender). These factors, along with broader economic inequality, residential segregation, and an often inadequate urban school system (Ainsworth-Darnell and Downey, 1998), work to reproduce the larger social order where whites have power and advantage. This white advantage and race bias is clear to minorities and often results in reduced hopes for success in a white controlled system (Ogbu, 1991; Persell, 1977). In fact, this is the important lesson learned from Ogbu's research (1978, 1991) on the reaction of youth (creating "oppositional cultures") to a clearly biased education system. Research showing the impact of race-biased educational systems on young minority student's achievement and attitudes is extensive (Cooper, 1996; Freeman, 1997).

Thus, it is not a distinct cultural orientation that works to the disadvantage of minority youth in U.S. schools. Rather, it is structured disadvantage that impacts the young person's hopes and achievements. African American youth are aware of this structured disadvantage in the educational system and in the broader society (Ogbu, 1978, 1985, 1991; Davis, 2004; Feagin et al., 1991). For many (especially in higher education), racism is a central component of their education experience (Feagin et al., 1991). Race stratified education systems contribute to distinct "white" and "black" cultures. It is problematic, according to

Fordham and Ogbu (1986) when "acting black" and "acting white" become identified in opposition to one another since "acting white" includes doing well in school. "Acting black" then becomes the opposite of that and thus implies not doing well in school. Recent summaries of scholarship on race and educational achievement in the United States stress this intersection between structural and individual (or cultural) components (Kao and Thompson, 2003). It should be noted that these race effects in education research are not merely class effects. Studies of youth from families with low socioeconomic status (SES) suggest that it is not low SES alone that affects the school experiences of minority youth. Both class and race work to the disadvantage of minority youth in the U.S. education system (Borman and Overman, 2004). Sometimes, however, the notion of agency comes into play for these youth. Some scholars have found that once socioeconomic background is controlled, African American youth are more likely than white youth to attend a college or university (especially those from low SES backgrounds) (Bennett and Xie, 2003).

The negative school experiences of many minority youth contribute to lower participation in higher education (Freeman, 1997). This lower participation is problematic for developing science talent among minorities. A number of dedicated researchers and educators have suggested educational policy and program strategies to work against and "resist" the race biases and inequalities in the U.S. educational system (Foster, 1993; Cummins, 1993; Haney, 1993; Ginwright, 2005). Some researchers have discovered responses to race discrimination that are positive and involve resistance rather then acceptance and lowering of achievement. Qualitative data suggest that some African American youth respond to discrimination by striving for academic success and their reports show that the motivation for this success is in the African American community (Sanders, 1997). Additionally, research on those minority youth that *do* succeed academically suggests that one of the keys to success is supportive school systems. Most notably, it is strong, supportive relationships with teachers that makes for the difference between low SES minority youth who succeed academically (even in the difficult mathematics and science curriculum) and those who do not (Borman and Overman, 2004).

There are a number of limitations in the work on race, gender, and science education systems. Although these limitations were reviewed earlier, it is important to reiterate the fact that research on African Americans in science, much less African American women in science, is scarce. It is impossible to conclude that all African Americans will have the same experiences in education and in science education. My focus here is on the experiences of African American women. In the pages below, I review the limited literature that examines aspects of race, gender and science.

In Chapter 2, I noted the strong interest in science shown by young African American women that was discovered in my earlier research (Hanson, 1996; Hanson and Palmer-Johnson, 2000). In this chapter on schools, it is particularly important to remind the reader of this finding, which is inconsistent with double jeopardy theory. It is consistent with other research that shows high interest in math and science among minority students in general. For example, Wenner's (2003) recent research found that poor, minority, inner-city elementary students had lower test scores in science than did their white counterparts but more positive attitudes about science and more time in science instruction.

In spite of the interest and involvement that young African American women show in science during the high school years (Hanson and Palmer-Johnson, 2000; Hanson, 2004), race discrimination continues to exist in the U.S. education system (both in general and in science) (Pearson, 1985; Swarat et al., 2004; Downey and Pribesh, 2004; Thomas, 1997; Felice, 1981; Feagin et al., 1996). Thus, young African American women experience special barriers in science education and occupations, and they often report a sense of feeling unwelcome in the science classroom.

Minorities, Schools, and Science Education

Before turning to a discussion of minority women in science education, it is important to note some of the issues that minorities, in general, face in the science classroom. Science is not just a male domain. It has historically been a white domain, as well. Early ideologies about natural inequalities by race influenced the work of scientists and scholars, as well as the treatment of minorities in the science domain (Smedley, 2002; Jackson, 2005; Pearson, 1978, 1982, 1985). Racism is a key feature of science in the United States and elsewhere. This has a large impact on the potential for success among minority students. Early work on science as fair has not been supported (Cole and Cole, 1967; Cole, 1987). Although underrepresented, African Americans have made major contributions to science (Ovelton Sammons, 1990). However, many are unaware of these contributions since teachers, curriculum, and textbooks often overlook minority scientists and their discoveries (Pearson, 1985).

Several trends have been pointed out in research on African American students in the science education system. It is important to note that the discouragement of African Americans in the elite arena of science has a long history in the United States and is not a new phenomenon (Pearson and Bechtel, 1989). Historically, educational policies in general have worked to ensure the advantages of whites (Bechtel, 1989).

Studies of African American scientists show that their parents often have high levels of education and professional careers. Like studies of white scientists, studies of black scientists show that education was highly valued in their family and community (Pearson, 1982, 1985). However, some have suggested that a parent's education may be less important than the preparation in math and science that students receive in elementary and secondary school (Grandy, 1998). African American students have lower performance on math and science standardized exams (relative to white and Asian American youth), and there is evidence of poor preparation in math and science during the early years (Pearson, 1991). Some have noted that minority students are overrepresented in low-income, inner-city areas where there is limited and poorly maintained science equipment. The personal, interactive teaching style that has been shown to encourage interest in science is less likely to be found under these conditions (Anderson, 1988). Additionally, a good number of African American scientists were trained at historically black colleges or universities (Pearson, 1982, 1985, 1998). These institutions are important in creating and developing science talent among African Americans. Unfortunately, these colleges and universities are often considered to be less prestigious and thus scientists trained there may not get the recognition they deserve (Pearson, 1987).

Another important finding in the research on race and science is that noncognitive factors, such as self-concept, an understanding of racism, and self-esteem, play a very critical role in young minority student's persistence in science education (Grandy, 1998; Rayman and Brett, 1995). Grandy (1998) argues that race discrimination literally impedes the cognitive development of high-ability minority students in science. Related to this, Steele (1997) found that many minority students' underperformance in school is not related so much to lack of preparation as it is to a "stereotype threat"; that is, minority students are aware of stereotypes about their intellectual abilities and are concerned with judgments in these areas. The concern may lead to a lowering of performance. Thus, the very pressure to disprove stereotypes about lower intellectual ability may result in anxieties and pressures that act to reduce performance (Swarat et al., 2004). A number of science programs that have been successful in retaining minority students have included components aimed at emphasizing confidence in the student's intellectual and academic abilities (Swarat et al., 2004). Rayman and Brett (1995) argue that self-concept is one of the most important factors in determining whether minorities (and women) select majors in science. Pearson's (1982) research suggests two factors at work. First, many black youth see science as too difficult and only available to the extremely talented. Second, many of these youth are systematically steered away (through tracking and counseling systems) from science.

Discrimination in recruitment and in encouragement are major factors that work to the disadvantage of minority groups in science classes and programs (Thomas, 1988). At the college level, institutional recruitment practices ensure that African Americans and minorities will be under-represented in college science programs. In the science classroom at all educational levels, race discrimination disempowers minority students. It negatively affects their cognitive development, as well as their affective development. The end result of the discrimination process is isolation of minority students from faculty, other students, and school organizations (Grandy, 1998). General school processes (especially at the high school level), such as discriminatory suspensions, expulsions, and misplacement in special education or low-ability classes (often because of biased standardized tests), also work to decrease the flow of talented African American youth in the science pipeline (Pearson, 1986).

One of the most important factors in discouraging minority youth in science involves teachers and teaching styles. Pearson (1986) notes that many white teachers have never seen African American scientists and engineers. African American students have few role models in science, since they may not have access to African American science teachers (or African American scientists in the curriculum). African American role models at school are especially important for African American youth given the fact that many do not have parents and relatives who are scientists (Thomas, 1988). There is much evidence that teachers often ignore African American students in the science classroom. This is even true for talented African American students (Malcom, 1983). Teachers do not believe that African Americans and non-Asian minorities are suitable candidates for science and ultimately discourage them (Pearson, 1986).

Additionally, there is evidence of an incompatibility between teacher styles in science classrooms and minority students' learning styles (Rosenthal, 1996). In situations where there is a mismatch between teaching styles and learning styles a student can study hard and still not get good grades (Rosenthal, 1996). It is important that math and science classes are taught in such a way that they incorporate cooperative, practical, laboratory, and applied learning with materials presented in multiple ways and limited time constraints (Rosenthal, 1996). As noted earlier, role modeling, self-confidence building, and providing a sense of belonging are also important for minority youth in the science classroom (Hoyte and Collett, 1993). Many of these strategies are not normal parts of high school or undergraduate science classes and programs. Yet these strategies, together with a diverse faculty and a school system that encourages faculty development on these issues, are key to promoting success for racial minorities in the science classroom (Hoyte and Collett, 1993).

Teachers have a tremendous impact on the career choice of young people and thus can be seen as part of the problem and the solution when it comes to recruitment of minorities into science (Pearson, 1985). Minority students report that their teachers are very influential in affecting their interest in science (Atwater et al., 1995). It is especially important that science teachers in middle and junior high schools present science curricula that stimulate the interest of all students, regardless of race/ethnicity (Pearson, 1985). Many African American students who were high achievers in elementary school science classes begin to show signs of low self-concept in science by the time they reach middle school (Jacobowitz, 1983). Additionally, some have suggested that African American youth have already made decisions about their future education and careers by the middle school years (Thomas, 1988). Rayman and Brett (1995) argue that the early middle and high school years are important in generating science interest and retention because once a student is in college, persistence rates often vary little by race (or gender). Pearson (1985) suggests that federal policies can do more to create and reward postsecondary programs that are successful in recruiting and retaining minorities in science programs. Additionally, he argues that the large number of African American students who are first-generation college students suggests a need for career counseling on opportunities for blacks and minorities in math and science-related fields (Pearson, 1991).

Minority Women, Schools, and Science Education

Studies of African American women in science document the considerable barriers that these women face (Carwell, 1977; Clewell and Anderson, 1991; Kenschaft, 1981; Malcom, 1976; Ovelton Sammons, 1990;

Rosamond, 1991; Vining Brown, 1994; Moses, 1989). The culture of science pervades the science classroom and this is a white, male culture that is often hostile to women and minorities (Harding, 1986; Rossiter, 1982; National Science Foundation, 2004). I have noted that students are provided little information on the contributions of African American scientists. Female African American scientists (their biographies and contributions) are even more neglected. Likewise, African American women are virtually absent in the research done by social scientists on those in the scientific professions (Pearson, 1985). Like African American scientists in general, these women are often trained at historically black colleges and universities (Pearson, 1985). Many of them end up working in academia (again, disproportionately at these historically black institutions).

Factors that help create a chilly climate for African American women in science overlap to some extent with those that affect all non-Asian racial/ethnic minorities. They involve teachers' negative stereotypes (and students' responses to them), lack of African American (and African American female) role models and mentors, and science content that excludes a focus on the contributions of women and African Americans. Kenschaft (1991) argues that some of these factors work to the detriment of majority women's experiences but many of them have stronger negative effects for minority (relative to majority) women.

Like the larger literature on minorities and schools, research on African American women's experiences in the science education system shows the critical role of teachers (Brown, 2000; Kenschaft, 1991). Unfortunately, African American women are often marginalized in the science classroom because of their race and gender. Science teachers tend to overlook these young women as a source of science talent. Malcom's (1976) research on African American women in science reveals low expectations of teachers, especially when it is a white teacher in an integrated school system. Clewell and Anderson's (1991) research on women of color in mathematics, science and engineering suggests that the greatest decline in attitudes toward science in this group is between the 6th and 7th grades. Some have suggested that the change toward a new learning environment involving male teachers contributes to gendered experiences in middle school (Catsambis, 1994). Those who do persist in science often see the role of teachers in their success (Clewell and Anderson, 1991; Matthews, 1984). In Jordan's study (1999) of black women in agronomic sciences, a large majority of them stated that teachers were an important factor in encouraging them in science. In Brown's qualitative study (2000) of African American women who had succeeded in math, one young women says the following about a white male teacher who did not pay her enough attention in her algebra class, "If I would have just been given the time or given a little extra push, I would not have made a C in the course" (Brown, 2000: 372).

Similar to the larger literature on minorities and education systems, there is evidence that the "stereotype threat" works against young minorities in the sciences (Swarat et al., 2004). In fact, the threat might be even greater in the sciences since this is the most elite area of the curriculum. Thus, young African American women's responses to negative stereotypes about the abilities of women and of African Americans in science might be a daunting and sometimes debilitating factor in their pursuit of science. Moses' study of postsecondary colleges and universities found that African American women in the science classroom often report having to face diminished self-confidence, cultural barriers, isolation, negative faculty evaluations, and a feeling of being underprepared. In spite of the large number of African American women entering higher education, they are still made to feel like "the other" and as if they are outsiders (Moses, 1989). Seldom are they encouraged to continue in postgraduate education (Moses, 1989). African American women are tested and re-tested against negative stereotypes.

Another important factor in the chilly science classroom climate involves the content of the science curriculum. Textbooks and teachers focus mainly on science knowledge and inventions created by white scientists. Hence, they are seldom made aware of the contributions of African Americans (much less African American women) in science (Ovalton Sammons, 1990; Von Sentima, 1985). Powell and Garcia (1985) and others have found few illustrations of people of color in science textbooks. They argue that it is difficult for underrepresented groups to see science positively and aspire to science professions when content and illustrations in science texts focus mainly on whites.

Another factor that creates a barrier and contributes to invisibility for young African American women in the science classroom involves the small number of minority women available as teachers, mentors, and classmates (Jordan, 1999; Moses, 1989). Colleges that have been the most successful in encouraging minority women scientists are women's colleges and historically black colleges and universities (HBCUs) (especially women's HBCUs), which have diverse faculties and give confidence to women (and minority women) in science (Jordan, 1999;

Pearson et al., 1999; Moses, 1989). Women's colleges that are not HBCUs may also play an important role in the production of African American women scientists (Leggon and Pearson, 1997). An analysis of the elite "Seven Sister" colleges suggests, however, that they historically exhibited some of the same race discrimination that was present in the larger society (Perkins, 1997),

Studies of African American women scientists have found that these women have experienced considerable amounts of racism and sexism (Pearson, 1985; Jordan, 1999; Moses, 1989). Like the general literature on African Americans in higher education, research suggests that African American women experience isolation, invisibility, hostility, indifference, and a lack of understanding in institutions of higher learning. Both faculty and students contribute to this environment in both overt and covert ways. It is not an environment that is conducive to academic development (Moses, 1989). Beliefs about "affirmative action" students work to further increase the hostility toward African American women in academia in general. These women may be stereotyped, resented, and treated with disrespect, since they are perceived as being less qualified. As such, they are largely treated as tokens and not as individuals (Moses, 1989). Interestingly, black women scientists suggest that it was racism that was the biggest problem in the educational system, while sexism was identified as the larger problem in the occupational system. Some women had a difficult time discerning which of these two sources of discrimination were working against them in certain situations. When race was reported as a problem, the nature of the discrimination varied from insensitive comments and acts to support staff that had trouble taking supervision from a black woman to colleagues that questioned their work. Others reported subtle discrimination as well as not-so-subtle efforts to isolate them and deny opportunities (Jordan, 1999).

Historical descriptions of African American women's attempts to enter math and science programs detail the roots of racism in these programs and help us to understand the continued presence of institutionalized racism and sexism. Rosamond (1991) describes the experiences of Vivienne Malone Mayes, one of the first black females to gain a Ph.D. in mathematics. In the 1950s, Vivienne Malone Mayes could not become a teaching assistant at the University of Texas in Austin because she was black. There were professors who would not allow black students in their classes. Talking about mathematics over coffee with classmates and advisors was not possible given the segregated cafes. She felt unwelcome and wrote the following about her first weeks in the program, "I was the only black and the only woman. For nine weeks, thirty or forty white men ignored me completely" (quoted in Rosamond, 1991: 39).

Research on school organization and practices suggests that these are influential factors in encouraging minority women in science. Some of the school factors that can make a difference in science persistence include encouragement and positive reinforcement, access to special programs, small classes, cooperative learning (group work), an active laboratory component, placement in advanced tracks, an early start in math and science courses, and participation in high school math and science clubs (Clewell and Anderson, 1991; Stempel et al., 2001). Like Berryman's (1983) argument in her classic report "Who Will Do Science?" I have argued in this volume that the negative school environment for women, minorities, and minority women in the sciences contributes to their low achievement (in spite of positive attitudes toward science). This low achievement is a critical factor in keeping minorities, women, and especially women minorities out of science degree programs and science occupations (Catsambis, 1995; Maple and Stage, 1991).

This chapter began with a reminder that African American women often have positive attitudes about science. The literature reviewed above shows considerable barriers for these young women in the school setting. I have summarized biased teachers, curriculum, and classroom interactions, as well as school practices that ultimately provide fewer opportunities for learning math and science to women, minorities, and minority women. Like others, Catsambis has concluded (1995) that low achievement in the early years and lack of learning opportunities are the major factors in keeping minorities and women out of math (and science). Research stressing the positive attitudes and interest in science among minority women (Hanson and Palmer-Johnson, 2000; Hanson, 1996) supports her research and ultimate conclusion that "[t]heir relatively high interest in mathematics could provide educators with an important opportunity to engage these students in more intensive academic efforts" (Catsambis,1994: 213).

I end the literature on African American women, schools, and science with a note on the role of social class in the science process. Race, class, and gender all interact to influence opportunities in the elite science education system (Hanson, 1996). Scientists are often white males from upper-SES backgrounds. The role of social class cannot be ignored when considering barriers to African Americans and African American women in science (Jordan, 1999; Pearson, 1986). Race continues to create an economic disadvantage for African Americans and, especially, African American children. Teachers, schools, and programs that are aware of class biases in the science education system are an important factor in developing all science talent. At the postsecondary level, Jordan (1999) suggests that faculty, advisors, and administrators can have a powerful impact by working to get greater financial assistance for these students, since finances are often one of the major obstacles to their persistence.

Findings from NELS

The literature reviewed above gives us a preliminary look at how schools affect young African American women's science experiences. Now I examine data from the Department of Education's representative, longitudinal NELS data set, as well as the data collected for this project to gain additional, more specific answers to the question of how school experiences affect science outcomes for these young African American women. First, I look at the analysis of the NELS data. NELS has excellent school measures (see Table A.4.1). Some of the variables refer to teachers (quality, interest, importance in science decision). Other variables measure the import of the counselor. Still others measure aspects of the school context involving, for example, school spirit, racial conflict, and school safety. Recognition and honors are also measured, as well as school program and school type (public vs. private). Our literature review suggests that all of these factors are important when considering the confluence of race, gender, and science in these minority women's experiences. I will begin with a description of young African American women on the school and teacher characteristics. We also compare the young African American women to young white women in order to determine how unique they are on these school and teacher experiences. Means are presented in Table A.4.1.

When it comes to teachers, the race advantage goes to the white women. These women are more likely to say that teaching is good at their school and that teachers are interested in students at their school. There is no race difference in response to the question about the importance of the teacher in the science decision. A rather small percent of women in both groups (32% of young African American women and 28% of young white women) report that their teacher is very important in their decisions about science. Note that the percent *is* larger in the African American group. Larger measurement error in this smaller sample may be partially responsible for the lack of significance in this race contrast. The young African American women are also more likely than their white counterparts to report that their counselor is important in their science decisions. In this case, the difference is significant.

Findings are mixed on the race differences in school climate variables. Young African American women are more likely to report that students are friendly to those in other race groups at their school. However, it is the young white women who report feeling more safe at their school. Finally, there is no race difference on the variable asking about school spirit. A majority of both groups feel that there is real school spirit at their school.

The young white women have an advantage on all variables measuring school honors and school programs. They are more likely than their African American counterparts to have won an academic honor, to have received recognition for good grades, to be in a private school, and to be in a college prep program at their school. Thus, it is the young white women who have a distinct advantage in receiving special recognition in the school system.

But how important are these school and teacher characteristics for young African American women's experiences in science education as measured in the NELS data? Table A.4.2 shows the school and teacher characteristics of young African American women who are engaged in science at various levels (e.g., in advanced courses, taking chemistry, choosing science as a major, etc.). Several school and teacher variables seem to be particularly important for distinguishing those who are and are not engaged in science education. These include quality of teaching at respondent's school, teacher's interest in students, winning an academic honor, receiving recognition for good grades, and high school program. As expected, in most cases the students who are more engaged in science have more of these school and teacher resources. Consider, for example, teacher interest in students. When teachers are interested, the young African American women are more engaged in science whether it be science achievement, course-taking (access) or attitudes. This pattern is particularly striking in the case of science attitudes. Here we see that, for example, 47% of the young women who report that teachers are interested in students are interested in science. The figure is 12% for those who do not report teacher interest. Large differences in the two groups of students also exist on science attitudes involving doing well in science, whether one needs science for a job, and whether one plans on a science occupation by the age of 30.

Another school characteristic "receiving recognition for good grades" is particularly important for distinguishing the young African American women on the science access and achievement variables. For example, 30% of those who had received recognition for grades had chemistry coursework in the 10th grade. Only 9% of those who had not received recognition for grades had this coursework. At least twice as many young African American women who had received recognition for good grades scored high on achievement variables involving science grades and standardized scores in various school years (relative to their peers who had not received recognition).

Finally, it should be noted that high school program was important in its association with engagement in science. Young African American women who are in college prep programs are often taking more science classes, experiencing higher science achievement, and professing more positive attitudes toward science. For example, 18% of the young African American women who were in college prep programs in high school reported a postsecondary degree in science. This figure was half as large (9%) for those who were not in college prep programs.

Another way to look at school effects is to combine the school variables in multivariate models that control on SES in predicting the odds of young African American women being engaged in various aspects of science. Results for this analysis are presented in Appendix Table A.4.3. Here we see that when the school variables are considered simultaneously, and SES is controlled, teachers continue to have important effects. Good teaching is especially important for predicting the odds of being in a science occupation in 2002. Teacher's interest in students has a positive impact on a number of aspects of course taking (science major and science degree) and attitudes (planned science occupation by age 30). Winning an academic honor and being in a college-prep high school program also increase the odds of success in science when other school variables and SES are taken into account. The largest effect in the model involves the effect of high school program on taking science courses. Odds of taking science courses during the past two years (when asked in 12th grade) increase by a factor of 25 when young African American women are in college preparation programs. Note that only one of the school variables " real school spirit at school" has no effect on any of the science outcomes in these analyses. Finally, it is important to note that when young African American women acquire these school and teacher resources, the outcomes most likely to be affected are science occupation (measured in 2000) and plans for a science occupation by the age of 30. Some of the effects in Table A.4.3 are anomalies. For example, feeling safe at school has mixed effects on science outcomes. This is a complex variable that might measure a variety of things, including whether or not the student is part of the in-group, and overall safety of the neighborhood where the school resides.

A consideration of school effects when other effects (family, community, and peer) are taken into account is also important for our overall understanding of the role of school variables in young African American women's science experiences. Logistic models examining these processes are included in Table A.4.4. Three school variables are included here: teacher's interest in students, academic honors, and high school program.[1] Results show that in this multivariate context, winning an academic honor is the most important school variable. This honor significantly increases the odds of success in science for all but one of the science outcomes (science degree expected by age 30). Teacher's interest in students is also a powerful predictor of science experiences in these models. The strongest school effect in Table A.4.4 involves the effect of teacher's interest on standardized science scores in the 12th grade. When young African American women have teachers who are interested in them, their odds of scoring in the top quartile on standardized science exams increase by a factor of 8 even when other important variables involving family, community, and peers are taken into account.

Findings from Knowledge Networks

Quantitative Analyses

Quantitative analysis of school effects in the Knowledge Networks data is shown in Table A.4.5. Here, a limited number of variables from each area of influence is included in a multiple classification analysis (MCA) model that shows the deviation from the mean on science outcomes for young African American women with various characteristics. Four school variables are considered here. They include several measures that are similar to NELS measures and several that diverge. School measures examined in Table A.4.5 include type of school (public/private), whether the school is co-ed, percent minority students in the school, and encouragement in science from adults at school. Results show that young African American women who are in private schools have an advantage in the area of science occupations. They are more likely to hope for, and expect a science occupation at age 30. However, the advantage on science grades and feeling good in science goes to young African American women who are in public schools. Interestingly, all girls' schools do not increase success in science for African

American women (as research finds for white women). In fact, participation in these schools has little effect on science outcomes but does
decrease the odds of expecting a science occupation at age 30. The
percent minority variable yields an interesting outcome. Young African
American women who are in schools where minorities represent less
then half of the student population are significantly more likely to hope
for (and expect) a science occupation at age 30. They are also more
likely to feel welcome in science. Finally, results show an important
effect of encouragement in science from adults at school (this could be
teachers, counselors, principles, etc.). Feeling encouraged by these
adults increases young African American women's high school science
grades, their feelings of being good in science, and their feelings of
being welcome in science. Some of these effects are considerable. For
example, even when other school effects, as well as family, community,
and peer effects are taken into account, the difference between those
who are encouraged by adults at school and those who are not is substantial. Seventy-three percent of those who are encouraged by adults
at their school feel welcome in science. The number is 52% for those
who do not experience this encouragement.

Thus, the survey data on young minority women's science experiences show that we cannot generalize from white samples. Past research
has neglected to focus on race and gender subgroups thus minimizing
our ability to understand processes for groups such as African American
women. Single-sex schools and other strategies that have been shown to
be effective for combined samples or samples of white women do not
necessarily convey to positive experiences for minority women.

Qualitative Analyses

As in other chapters, I supplement the quantitative analyses of the
NELS and Knowledge Networks school data with an examination of
the qualitative, open-ended responses that the young African American
women provided to queries about their science experiences involving
teachers and schools.

Some of the qualitative feedback came in response to the young
girl's situation in the vignette. Other feedback came in responses to
questions about why young African American women are or are not
interested in (and encouraged) in science. Some of the qualitative
answers came in response to a specific question about whether (and
how) teachers and counselors encourage them (the respondent) in science. Finally, the young women were asked how they would change

school systems so that all students (regardless of race or gender) were interested in science and felt good about their science abilities.

The themes that I discovered in the qualitative responses reflect many of the topics that were revealed in the review of the literature on schools, minorities, gender, and science. A large number of the responses reflected problems in school systems and classrooms that take away from interest in science, for example, discrimination, poor teaching, and the dearth of minority teachers. Some of the responses focused on issues having to do with the young African American woman. For example, the fact that *science was too hard* and they *lacked preparation* were somewhat common responses. Issues mentioned focused on the following factors:

- Science is too hard
- Lack of preparation and limited classroom (or school) resources
- Minorities (especially African Americans) and women are absent in the science classroom and science content (books, lectures, etc.)
- Being ignored
- Race (of teacher or self) as a negative factor
- Isolation in the classroom
- General classroom environment
- Discrimination
- Poor teaching
- Lack of encouragement (or discouragement)

In the following paragraphs, I provide some examples of the young women's statements in each of these areas beginning with the ones focusing on the young girls' sense that science is too hard and they lack preparation and then continuing on to show comments about school and classroom experiences. Although the comments (e.g., involving isolation and discrimination) are, on the whole, reflections on the difficulties for the African American women in science education, the section ends with a positive note. I present some of the statements made by the young women that reveal their agency in the science education system. I also include some of the young women's suggestions for changes in educational institutions that would provide more encouragement in the sciences for young women like themselves. As one reads these young women's reports of science experiences and frustrations, the "tide" that they are swimming against becomes clear, as does their strength in attempting to stay afloat.

Science Is Too Hard

The fact that many African American women are discouraged in science because of the difficulty of the field was discussed in an early chapter on African American women's experiences in science. This difficulty is also related to the science classroom and the environment there. One young woman reported that science "seemed like an area of education for only a select few." This sense of an elite environment was reflected in other young woman's responses such as, "I felt that I could never do as good as I hoped because the teacher was talking on a Ph.D. level and I felt very lost" and "science ... is a very competitive field ... it makes it intimidating." Other young women expanded on the consequences of this intimidation for people like themselves: "They are just scared that they can't do it. They think the work will get so hard that they will not be able to do it." and "Sometimes students might not understand the concepts, so they might feel rejected."

Lack of Preparation and Limited Classroom (or School) Resources

One of the most poignant responses that reflected the limited resources and lack of preparation for many young African American women came in this young woman's response: "Science is in our hearts as black people. We just need to be given the help in school. The right tools just like everyone else." Others suggest that African Americans are not given "enough basics" in their education and do not end up with the same "educational foundation" as other students.

Some of the young women were very specific about the lack of resources for science education in their schools. They refer to black high schools that are "not teaching science like it should be taught," and predominantly "African American community ... school[s] not as equipped as those in other areas," with few "hands on opportunities" which are needed in order to "understand and love science." The economic source of these problems is noted in one young woman's reply: "I guess ... most African Americans grew up in inner city schools [that] did not receive the funding they are supposed to receive."

The consequences of this lack of preparation for science come through clearly in one girl's description of her experiences in the science classroom: "When we study the inside of the body in biology, we get to do a project that most all of the minority students don't know what to do and are in line for help. All the white kids are all done with their projects by the time we start."

Minorities (Especially African Americans) and Women Are Absent in the Science Classroom and Science Content

The young women in the survey often comment on the lack of African American teachers and the absence of African American scientists in their science lectures and texts. This observation is sometimes linked to their feeling of being out of place in science. As one young woman wrote, "They really don't talk about black scientists unless it's black history month, and if you go to a school where there's mostly whites in your class how are you supposed to feel like you belong?" Another young woman shares this sentiment, "Sometimes I do feel out of place.... I ... see only white scientists in my textbooks and in the films we watch." This absence of African American scientists is noted even in minority schools, as suggested by one young woman, "Our school is 85% [minority] and there is still no black man or woman to look at in these books. I feel not so free." One young woman was discouraged by these trends; she writes, "African Americans have a hard time trying to get their foot in the door of anything positive and self-beneficial. Science is a white dominated field and there's a lack of highly visible role models in the field for African Americans." Finally, the lack of African American role models in the science classroom was connected to lack of success in science by one young woman, "We don't have role models for our people. You're more geared to learn and do good in school if you have someone like you there to teach you what you need to know."

The lack of women scientists in the classroom and textbooks was also noted by some respondents. One young woman notes, "There's Einstein, Newton, etc., that constantly get [noted], I hardly remember hearing about more than two female scientists in schools." Another respondent cannot remember hearing about any female scientists: "There's not enough recognized female scientists in high school study books. I do not recall any study of female scientists when I was in high school. I only remember studying male scientists and inventors. Even though, there are plenty of [women] working or studying science I believe that men are mostly recognized." Thus, this young woman attributes the limited focus on female scientists to bias, not to the absence of female scientists.

Being Ignored

Some of the young women feel ignored in the science classroom. One respondent puts it quite simply, "I had two biology professors who would not even acknowledge my presence in class, except to tell me I was going to fail." Another says this about her later (college) experiences in the science classroom, "I was practically invisible to several of my professors." Being ignored lowers confidence, as reported by one young woman, "I didn't get called on in most of my classes even if my hand was raised first. My teachers didn't give me confidence that I needed so I stopped raising my hand." Thus, the women are making efforts but sometimes not getting encouragement. For some, there are reports of a negative response: "I really try hard but the teacher never acknowledges me when I raise my hand. And when I ask for help she has this harsh attitude which makes me mad." Similarly, one young woman adds, "My ... teacher doesn't like me. When I ask questions, he acts like he doesn't want to answer them. He also never picks me when I raise my hand."

The young women's reports of being ignored in science classes are sometimes discussed in the context of gender. They talk about being "overlooked in favor of male students," of the teachers paying "more attention to the males" and "being more willing to help males when problems arise." One young woman provides a long list of the math and science classes that she had experienced this in, "I have been over-looked for a male to respond in math and science classes. This has happened in calculus, biology, chemistry, and physics." Sometimes the feeling was not so much that the teacher was pro-male but anti-female. One young woman suggests, "The teacher did not care if the female students understood or even if we studied or did our homework." In the next section, some of the issues that are related to race in the class-room are discussed.

Race (of Teacher or Self) as a Negative Factor

The importance of having African American teachers in the classroom is highlighted vividly in one young woman's description of her experiences, "There are a lot of African Americans out there that love science. In my school, the majority of the teachers were white. They didn't give us the confidence that we needed. We lost that great feeling about science and try to do something else in another area, where we are looked at. That is why I started in business, most of the business classes were taught by the same African American teacher and she gave me that confidence about business." The most powerful phrase in this young woman's reply involves wanting to study in an academic area where "we are looked at."

Many young women report on science classrooms in which white teachers teach about white scientists. One young woman states, "Most of the science classes were taught by white teachers. I felt like they looked at us like we [weren't] supposed to be scientists." Similarly, one young woman writes, "Most teachers do not think we are smart enough to be in science," and another writes, "Don't think he [the science teacher] likes black people." This classroom environment lingers into later science experiences, as suggested by one young woman, "Even now in medicine … the disparity is huge. Most of my medical school teachers were male and white male at that. There were no black minorities. Only foreign minorities were represented.… I've gotten my degree and I still don't feel welcomed."

Isolation in the Classroom

The absence of African American teachers, the limited focus on African American scientists, and the sense of being ignored (and not supported) lead to a feeling of isolation in the science classroom for many of the young African American women in the survey. The young women who felt particularly isolated were those who were in primarily white educational environments. One respondent connects this isolation to feelings about belonging in science, "I went to a high school that was predominantly white and that can make an African American feel out of place. It can make you feel as though this particular subject is not for us."

Even when the young women are in integrated schools, the science classroom is sometimes dominated by white students and white teachers. One young woman responds to this situation after hearing about LaToya's experience in the vignette, "I felt the same when in my science classes in high school. In most of my classes, I was the only African American, and that made me feel uncomfortable." Another girl also emphasizes, "There were very few, if any other black students taking science classes in that field when I was in school. Also, science is often considered a difficult subject and it sometimes feels that mainstream culture has no expectations for us to achieve in that area." Thus, the young women's reports remind us again and again, that they do not feel comfortable or welcome in the science classroom. In fact, the young women often do not feel as if their teachers and the larger system have very high expectations for them at all in the realm of science.

General Classroom Environment

The young women's statements about the low expectations held by teachers and school systems (and society as a whole) suggest the general environment in the science classroom is not always a positive one for them. The sense of isolation that some report is related to a broader feeling of a classroom environment that is not supportive. One young woman sums it up with her comment, "Well if you can't relate to anyone, and you don't want anyone to cater to you [this is a problem]. Science class often [involves] groups and your white counterpart feels like you're looking for a ride ... does all the work ... in short your ideas aren't considered."

Race is often a subtext for this uncomfortable environment in the science classroom. In fact, one young girl suggests that "there are not enough black students taking an interest in science" and another describes what is needed to feel comfortable when she says, "Sometimes it ... has a lot to do with the percentage of blacks in the school itself as to how comfortable an African American may feel in class." Thus, for another, "If you go to a school where there are mostly whites in your class how are you supposed to feel? Like you don't belong."

Discrimination

Sometimes it is not so much isolation and uncomfortable classroom environments as it is real and active discrimination that the young women report experiencing. Both race and sex discrimination were mentioned but more young women cite race discrimination as a factor (or a more serious factor) than is the case for sex discrimination. The following comment clearly describes the young woman's sense of race discrimination in the science classroom, "Because of the atmosphere you get in the science classroom, I noticed if I needed help on a project and a white student needed help also they would give it to the white student. The white student was considered to be more interested." Other comments include:

> "We are the minority in science and most teachers do not see us as excelling ... and therefore do not take the time needed with us."
> "There are [racial] prejudices still found in a lot of the country."
> "People still look at African Americans as not being smart enough, and not being sincere."

"White people always try to make it seem like they are better than black people in everything not just science."

"If you're an African American, people don't see you the same as whites."

"I think blacks have never been encouraged to pursue medicine or science. This dates all the way back to slavery."

"It has always been harder for an African American to obtain their goals."

"The odds are stacked against us."

"Some people doubt that we can do it (science)."

"Blacks have a harder time in everything we do."

"Blacks in general are not seen as the science type."

"They are seen as more physically inclined, i.e., sports, than cerebrally."

"People still have the mindset that African Americans are less than human and can't make it in a white world."

The consequences for later science achievement and occupations are reflected in one respondent's comment, "The white race does not accept African Americans as a whole.... It would be exceptionally hard for white America to accept an African American as a lead scientist." The young women do not see this environment as one that rewards their efforts. One young woman notes, "I have had to work twice as hard as my European counterparts to earn less." Unfortunately, the women see this racism as endemic in the school system, "The American school system does not have African American children's interest as a big priority, unless it will benefit an American interest." These statements speak to the "tide" that so many young, talented African American women are swimming against when they enter the realm of science.

The young women also see sex discrimination in the science classroom. Again, as with race, there are specific classroom experiences that the young women remember as being sexist. One young woman recalls, "Whenever we would have a science experiment, the teachers would only pick the boys to showcase their results, and not even reprimanding the girls if they didn't complete their assignments." As with the race stereotypes noted above (African Americans are less "human," less "cerebral"), some of the comments revealed stereotypes about women. One young woman suggests, "Men in science, teachers, etc., … think a woman cannot think for herself to get the job done, some probably think we have brains the size of a pea!!!" Another comments, "Some people say … girls can't do this, they can't do that." The young women see the consequences of the stereotypes and their pervasiveness. One young woman comments, "Most of the male science teachers I've ever had tend to call on boys more. I think what she [LaToya] is going through is common in many schools today." The young women do not agree with these stereotypes, as suggested by one young woman's comment, "Many male teachers still operate under the misconception that boys are smarter in math and science than girls."

Finally, some young women noted both racism and sexism in their comments about their science experiences. In response to a vignette in which a young black woman, LaToya, feels uncomfortable in the science classroom, one young woman is very empathetic when reflecting on her own experiences:

> "I came away with a real feeling of contempt. I can think of any number of specific reasons LaToya might have for not feeling as enthusiastic about science as she once did, but that vague feeling of not belonging I felt was designed to irritate me. Due to the sexism and racism that is part of our world, and how that is manifested in the classroom, it does not take much for me to think that LaToya could be treated any sort of way in her science classes."

Some young women actually use the term "discrimination" when voicing their thoughts, "I think that LaToya is in a male dominated environment, I feel that the teacher is discriminating against her. Due to her sex and the color of her skin."

Poor Teaching

One of the factors that is often noted in the young African American women's reports of science classroom experiences involves poor teaching. Even when there is not a sense of being ignored, isolated or discriminated against, many of they young women struggle with teachers that are not successful in making science interesting and exciting. Reports of teachers who "can't teach," "are boring," "don't know how to teach [science] to others," "never told you how much you would use it in the real world," "unwilling to help you really understand the information," "didn't really explain it well," "didn't make it interesting" are common. One young woman concludes, "I have never had a teacher be excited about teaching it." These experiences can result in more than a lack of interest in science. They can create an antipathy toward science as suggested by this young woman's remarks: "One year I even had a teacher fresh out of university. He came in the class, never bothered to learn names and just started writing stuff on the board like he would for an advanced college course. That was very frustrating. It also made me have ill feelings towards how science would be in college once I got there." More specifically, one young woman refers to a combination of poor teaching and being ignored, when she reports the "uninteresting things included rote memorization of a pile of notes that were hard to understand in the first place, being ignored when I was confused, and impatience of teachers who were not willing to slow down to show me what to do."

As with many of the other comments, a good number of the comments about teaching involve the notion of race. One young woman suggests that "African Americans ... might not understand that much or the teacher doesn't explain enough information." Another young woman speaks about the lack of focus on science in majority minority schools, "I don't think that African American schools encourage science ... so we would have a tough time learning." The sense of hopelessness that arises when schools and teachers are not interested in them is clear in this young woman's response, "Nobody cared. I think to the majority of my teachers, it was just a job for them to go to everyday."

Lack of Encouragement (or Discouragement)

Finally, sometimes the young women don't report negative experiences (like being ignored or discriminated against) so much as just a lack of encouragement. One young woman "cannot recall a time when a teacher or counselor specifically told me to pursue science." Another recalls that, "The teachers were mostly just satisfied with fair grades. They never tried to encourage me more." Still another respondent states simply, "No role models, no expectations for achievement." Another concludes, "They didn't care." As the one young woman's comment (above) suggests, neither teachers nor counselors were seen as sources of encouragement in science. Another girl talks specifically about her counselor, "My counselor really didn't encourage me into anything.... There were entirely too many students and too [few] counselors."

The lack of encouragement seemed pervasive in some cases, "I just haven't seen or heard anything in my school from any adult that would encourage me in the field of science." This lack of encouragement is discouraging as one respondent notes, "No one tells me that I can do it so I believe they think I am not capable of it. I feel discouraged [in science] because they don't think I can do it."

Sometimes, the young girls experienced active discouragement in science. One respondent recalls that "the negative feedback I got especially by my teachers … [was] a great influence on my losing interest in science."

Agency

A good number of the young African American women responded to the vignette about LaToya's difficulty in the science classroom with encouragement and support. Often these were the same women that reported feeling uncomfortable in the science classroom (as did LaToya) as well as feeling isolated and discriminated against. One young woman's advice was, "Don't let anyone lead you to believe that your sex or race will keep you from going to the top because there's absolutely nothing anyone can do to you or say to make you feel inferior without your permission." Others agreed that LaToya should not let the opinions of others affect her. One suggests, "Why should it matter if she fits in the class? She should be more concerned with her grades and test scores. If she is receiving the same type of grades in her class then it really shouldn't matter." It is clear (from the respondents' reports) that these messages and sources of agency are not ones that come from the school environment. Rather they come from the family and community support that is discussed in the next chapter. One young woman concludes, "I don't believe anyone can make you feel unwelcome from what you love."

Some young women gave LaToya specific suggestions for the classroom. One suggests, "I feel she should stand up to her teacher and ask why she isn't called on as much as the other students." A good number of the respondents agreed that LaToya should speak with her teacher and let him know how she feels. Others suggested, "Be louder with answering questions. Make sure that your teacher sees your hand up. Interact with your classmates so they will know you."

Suggestions for Change

The research reported here has shown schools and teachers to be critical for success in science. Yet both the quantitative and qualitative data show that the young African American women experience significant problems in the science classroom. When asked what they would change, the answers centered on issues involving teaching, diversity, equal treatment, better labs and resources, and special programs. Samples of the respondents' comments are provided below.

Teaching

"First of all a teacher's excitement and interest has a lot of weight on how a child would … learn. A teacher should make every student feel as important as the next."

"You can only change so much but I would definitely monitor the teachers."

"I would separate the men from the women, but they would both be taught the same curriculum."

"Field trips, show and tell, hands on, and many of other sorts of demonstrations."

"Make it interesting, not boring."

"Give more insight on where a science degree can take them. What careers they could have and how much money they could make."

"Don't make it out to be hard and try to help students understand the information in a fun way."

Diversity

"I would change all the books and have lots of people of color [in them]."

"Bringing more diversity among educators, and set a specific unbiased science curriculum that will help identify potential in students regardless of their race or gender."

"More minority and female teachers."

"Get more women to come in and be role models."

"Change the topics of the science class to demonstrate more diverse issues in the class."

"I would make the staff more diverse and try to integrate schools that are doing well with those who are not, to even out the playing field."

"More African American teachers."

Equal Treatment

"I think everyone just needs an equal chance to participate."

"I would just tell whoever to give everyone an equal chance because I am a Jamaican American and they single me out and say that I can't do it so I just want to be able to have an equal chance like everyone else."

"Teachers need to show an interest in ALL of their students."

Better Labs and Resources

"I would urge the school system to provide better labs and equipment."

Special Programs

"Establish programs and centers to educate minorities and women and children about science."

Conclusions

The myth that schools are an equalizing force in a country where every child has equal opportunity for success hides the reality of inequality in educational access, resources, and opportunities. Young African American women enter education systems that have race and gender biases. Historical descriptions of African American women's attempts to enter math and science programs detail the roots of racism in these programs and help us to understand the continued presence of institutionalized racism and sexism (Rosamond, 1991). Research suggests that minority youth (and minority females) are aware of these biases (Ogbu, 1991; Davis, 2004; Olsen, 1996). Inequity in educational systems often reduces the achievement of minorities (and female minorities) but positive attitudes toward school (and toward science) are more immune (Hanson, 1996; Maple and Stage, 1991). Studies also show that those who do succeed academically often find the keys to success in supportive school systems (Borman and Overman, 2004). Increasingly, data from the National Science Foundation (2000, 2004) has revealed a strong presence of African American women in post-secondary science programs and in science occupations. Findings from the analysis of the NELS and Knowledge Networks data sets lend additional insights into the complexity of school effects and into the ways in which teachers and schools encourage and discourage young African American women in the science domain.

Findings from the NELS data reveal that on a good number of school characteristics involving teachers, school honors, and school programs, the race advantage goes to the young white women. Multivariate analyses reveal that these factors are key explanations for engagement in science education. Winning an academic honor and having teachers who are interested in the student are two of the most cogent school effects in the analysis of science experiences. African American females in the sample have a distinct disadvantage on each of these school resources.

The young African American women's responses to open-ended questions in the Knowledge Networks survey provide a wealth of information about their experiences in the school system and in the science classroom. Although the young African American women are often interested in science, their remarks show the difficulties they experience when they enter the school system. The next chapter shows considerable support that families provide for this interest in science. No evidence of a distinct cultural orientation that works to the disadvantage of these young women is found. Rather, it is structured disadvantage that impacts their hopes and achievements. Like other research on minorities, mine shows that young African American women are aware of this structured disadvantage in the educational system (and in the broader society) (Ogbu, 1978, 1985, 1991; Davis, 2004; Feagin et al., 1991). Like Feagin et al.'s (1991) research on college experiences, I show that for many of the young African American women in the sample, racism is a central component of their science education experience.

Web survey findings show that although the young women are often discouraged by lack of preparation, isolation, lack of diversity, discrimination, and poor teaching, they also have considerable agency when it comes to their experiences with science. Importantly, they are observing and thinking about their experiences and are able to reflect on the nature of the science classroom, how it affects them, and how it can be improved. Their thoughts often provide insights into the way in which schools and teachers veer them away from science and lower their confidence. They also provide insights into how schools can change and the need for an equal footing in science.

I turn now to a look at the literature examining race, gender, and family (as well as community) influence on science experiences.

5 Influences—Family and Community

"My Mother Never Minded Me Using Her Kitchen Utensils to Dig Up Insects and Worms to Explore."

My analyses of school systems suggests that young African American women often get little encouragement in science from the adults that they encounter there. Schools systems involve race and gender structures that are particularly cogent in the elite science system. Nevertheless, I have discovered a high level of interest and involvement in science among young African American women. I turn now to an examination of a potential source of agency for these young women—the African American family and community. Research has suggested that African American subcultures provide young women with a unique set of resources—resources that might be important for generating interest and success in science (Hanson and Palmer-Johnson, 2000). These are essentially resources for resistance given the chilly climate for African American women in the science classroom.

This chapter begins with a brief look at the literature on African American family, community, and resources for young women in science. The emphasis is on the family and community as a source of agency and social capital for young African American women in science. I then look at the NELS data to quantify family and community characteristics and their potential influences on science education experiences in a nationally representative sample of young African American women. I also use the more recent Knowledge Networks survey data to quantify family influences on science experiences. Finally, I turn to the qualitative Knowledge Networks data to gain further insight into young African American women's perceptions of their family and community resources and influences. The quantitative data may be important in measuring traditional family and community factors and the level of significance attached to measures of their effect on science outcomes. However, the qualitative data from Knowledge Networks and the young women's own words will provide insight into the complexities and nuances of family and community experiences that will be as helpful as the quantitative data in understanding young African American women's interest and involvement in science.

Background: The Literature on African American Family and Community Influences

Research on minority families often has a less than positive tone. Early work by Moynihan (1965) gave credence to the notion of the deficient minority family. There is much evidence that this notion is alive today (Eitzen and Zinn, 2004; Eshleman, 2002). Early work by Billingsley (1968) and later work by McCubben et al. (1998) and others (e.g., Roschelle, 1999) questioned this negative stereotype and showed considerable strength and resilience in the African American family. My earlier work (Hanson and Palmer-Johnson, 2000), and the approach taken in this book, also support this notion of strength and agency by showing the positive family influence and encouragement that work to support young African American women in the science realm.

It is true that many minority families have fewer of the socioeconomic resources that positively influence young people's success in science (and education in general). Related to this, school systems and educators often have a less then positive view of the minority family (Liontos, 1991). Consequently, minority families are less likely to be viewed as resources that might work together with schools for the benefit of young people. The fact is, research on minority families suggest that they have high educational expectations for their children and high involvement in their lives. Minority families provide considerable encouragement to their daughters, and they offer them a wealth of resources including a history of strong, working women who combine family roles with work roles and an emphasis on independence and education (not marriage) as a source of mobility. Some have suggested that the high positive self-concept found among minority youth under conditions of economic and social oppression is based in distinctive minority community, church, and family groups that have historically provided a unique support system and encouraged children to be positive and proud (Luster and McAdoo, 1995; Foster and Perry, 1982). The approach that many have taken in understanding non-economic sources of support that come from community networks is that of social capital.

Increasingly, social scientists are using the concept of social (and community) capital to understand the consequences of social networks and ties for a broad range of positive individual and community level outcomes. Social capital consists of social organizations involving networks with normative systems that facilitate coordination and cooperation for mutual benefit. Put more simply, social capital consists of resources that individuals can draw upon based on their social ties and relationships with others. Here, the social capital that young African American women gain in the family and in the larger African American community (including the religious community) is particularly interesting. Political scientist Robert Putnam (2000) argues that social capital is used to increase social cohesion, resolve collective problems through individuals organizing for change, and increase the sharing of (and access to other) resources. Some (Warren et al., 2001) have argued that social capital can be used to counter the lack of economic capital in poor communities. Not all African American families and communities are poor. But the history of racism and discrimination in the United States has contributed to ongoing race inequality in education and income.

The African American family and community provide a type of social capital (community capital) that contributes to young women's success in science. The networks and norms that typify African American families and communities are unique and perhaps uniquely suited for conversion into science success. Additionally, some norms, such as the stress on giving back to the community, provide particular fuel to succeed in the high status science field. Thus, in the African American family and broader community where networks work together for support and sacrifice is common, individuals feel obligated to give back to the community. As the literature below suggests, today's African American communities are making particular investments in young women and thus young women, especially, might feel obligated to give back.

Historically, African American women have always worked, and the African American family has exhibited greater gender equality in family decision making and division of labor than in the larger society (Gutman, 1976; Hill, 1971; Kane, 2000). In the cultural context of the African American community, women who worked and who headed families were not atypical (Andersen, 1997). African American women continue to have high rates of labor force participation and are more likely than white women to be head of the household. Part of the explanation for the recent high rates of labor force participation among African American women comes from the decline in employment opportunities for African American men and increases in employment opportunities for African American women that began after World War II (Spain and Bianchi, 1996). There is substantial evidence that a mother's educational and occupational resources have an impact on daughter's behaviors and aspirations (Mau et al., 1995; Rayman and Brett, 1993). This influence may be even stronger in the African American family than in the white family (Maple and Stage, 1991).

The historically high rate of labor force participation among African American women has contributed to unique gender ideologies. Work and family roles are not in conflict, as is often the case in the white community (Collins, 1990b). Collins (1990b) argues that African American women have been integrating their economic-provider activities and mothering activities since the days of slavery. Instead of work being in opposition to motherhood, work is seen as an important dimension of motherhood. For many young white women, it is the perceived incompatibility between science careers and family pursuits that keeps them from entering and persisting in the science pipeline (Matyas, 1986; Ware and Lee, 1988).

African American families (more than white families) emphasize education and occupations as sources of mobility for their daughters (Higginbotham and Weber, 1992). The above patterns have contributed to more androgynous gender roles and greater self-esteem, independence, and assertiveness as well as high educational and occupational expectations among young African American women (and their parents) relative to other women (Andersen, 1997; Hanson and Palmer-Johnson, 2000; Hill and Sprague, 1999; National Center for Education Statistics, 2000a). All of these characteristic are related to success in science (Hanson, 1996).

One of the ways that the stress on giving back to the community might be evidenced, is in community and church involvement and volunteer activities. These activities can be seen as collective actions that contribute to the collective good (Musick et al., 2000). Increasingly, researchers have found that volunteer and community work is associated with a wide range of positive outcomes for the individual, including academic outcomes (Wilson and Musick, 1997; McNeal, 1999). Work in the community cannot be fully understood without taking race and gender into account. Recent research by Cidade (2004) shows that young white girls might be more likely than young African American girls to answer in the affirmative on general questions about volunteerism, but it is African American girls who report more overall time volunteering.

Higginbotham and Weber (1992) argue that African American women are more likely than are others to connect their upward mobility to a racial uplift process. Their success is not seen as an individual activity but rather one that involves the support of family, friends, and community. African American women, more then others, might feel a sense of societal debt since their success will be (or has been) a result of a group effort and not just individual work. Higginbotham and Weber conclude that upwardly mobile African American women are almost twice as likely as upwardly mobile white women to feel a sense of debt to family. It is important to keep in mind the uniquely high investments in daughters in the African American community (noted above). African American communities have been shown to provide more support and opportunities for growth to their daughters (since they are better investments) than to their sons. Thus, these young women might feel a particular obligation to give back to the initial community, whether it is the actual family and physical community that the individual was raised in or the broader community. Thus the ideas of community capital, success, volunteerism (or giving back), race, and gender come together in a unique way in the African American community. The combination of factors is one, I argue, that might contribute to a desire to succeed in the most elite area of the educational and occupational systems—science. The stress on giving back to the community and success in science as a way to give back is, I argue, a unique aspect of the African American system that needs to be considered in this study of African American women in science.

Findings from NELS

What do the analyses of NELS data collected by the National Center for Education Statistics on a nationally representative sample of high school youth tell us about family and community influences on young African American women's science experiences? Let us begin with a look at the family.

Family Characteristics and Effects on Science

This section begins with a description of young African American women on a number of general family characteristics. It also compares the young African American women to young white women in order to determine how unique they are on these family variables. Means presented in Table A.5.1 show that young African American and white women diverge on a majority of the family characteristics examined. The first thing to note is that, as one might expect, the young white women score higher on a majority of the family variables measuring educational and occupational aspects of family socio-economic status. For example, the white women are twice as likely to have a mother with a college education and to have a father in a professional occupation. Surprisingly, however, there is no difference between the two groups of young women on whether their *mother* works in a professional occupation. Thus, young African American women do not have a disadvantage on this factor that has been shown to be an influence on young women's interest in non-traditional occupations such as science.

What about non-socioeconomic family characteristics involving parent's and others' investments in their children? Responses to questions asking about mother's and father's educational expectations for their child, discussions with parents on school-related activities and importance of family members in science decisions are also provided in Table A.5.1. The race difference is statistically significant on a majority of these measures *and in virtually every case where there is a difference, it is the African American women that score higher.* Young African American women are more likely than are their white counterparts to report that their mothers and fathers have high educational expectations for them. They also are more likely to report that their mother and close relatives desire them to attend college, that they discuss grades with their parents, and that parents and siblings are important in their science decisions. Some of the largest race differences occur on the variables measuring importance of parents and siblings in science decisions. In the case of siblings, for example, the young African American women are almost three times as likely as are the young white women to report that their siblings were an important influence when making decisions about science (21% vs. 8%).

How important are these family characteristics for young African American women's experiences in science education? To answer this question, I first look at the science experiences of African American women in different statuses on the family variables. I then perform a multivariate logistic analysis in which we include the family variables (as well as a general control for socioeconomic status) in order to predict the odds of doing well on the various science measures at different points in time.

Figures in Table A.5.2 show means on science outcomes (and tests for significance of difference between groups) for African American women in various family statuses. The important thing to note about the information presented in this table is the large number of science outcomes (involving access, achievement, and attitudes from the 8th grade through the post–high-school years) that differ significantly across family statuses. That is, young African American women who vary on characteristics such as mother's and father's education, mother's and father's educational expectations for the respondent, etc. often vary on their science experiences. When one looks at a family variable and examines its association (across the row) with various science experiences, it becomes clear that each of the family characteristics is associated with differences in science outcomes. The family characteristics that are associated with the greatest variation in access to science include mother's and father's education, importance of siblings in science decisions, and father's (and close relatives') desire for respondent to attend college after high school. In most cases (but not all), those who had parents (and other family members) with higher education, educational hopes, and engagement were more likely to do well in science. This is especially the case for the variables measuring mother's and father's education. For example, 35% of those whose mother had a college education had majored in science in postsecondary school, but only 23% of those whose mother did not have a college degree majored in science in postsecondary school.

Family variables have an even higher association with science achievement in the sample of African American women. In some cases (e.g., father's education and importance of siblings in science decision), the family variable is significantly associated with every one of the science achievement outcomes. Interestingly, having a father with a higher education is associated with more positive science achievement on all but the occupation variable. Conversely, having siblings that are important in science decisions is inversely associated with science achievement on all but the occupation variable. Here, it is siblings, not fathers that are more important in understanding which women are in science occupations. Most of the family variables are such that those who are higher on the variable (e.g., mothers and fathers with higher educational expectations) have higher science achievement. For example when young women have a mother who has in a professional occupation they are over three times as likely to score in the top quartile on standardized science exams in 8th and 12th grade (28% vs. 8% in the 8th grade, and 16% vs. 5% in the 12th grade).

The family variables that appear to be most associated with positive science attitudes for the young African American women are the ones measuring investment, not socioeconomic status. For example, father's education is not associated with very many of the science attitude outcomes but father's desire for respondent to attend college after high school is associated with a large majority of the science attitude outcomes. Again, in most cases when family members have higher hopes and are involved in discussions and decisions with the respondent, the respondent has more positive attitudes about science. For example, more than half (51%) of the young African American women who often discuss their grades with parents reported that they were interested in science in the 12th grade, while approximately one-third (34%) of those who sometimes or never discussed grades with their parent reported this interest.

An examination of how the family variables influence the science outcomes when considered together and with a control on family socioeconomic status is provided in Table A.5.3. Given the number of variables, size of sample, and requirements for logistic solutions, all of the family variables in the model could not be included; however, a majority are included. Since these are multivariate causal models, and the family variables were measured in the senior year of high school (1992), only science outcomes for the 12th grade and beyond are included. It should be noted that a number of family and science variables have missing values that, when combined in a multivariate model, produce a small N for the analysis (especially given the fact that the original sample of minority women is already small). Thus, I suggest these results be considered as exploratory. Findings here show that a majority of the family variables have at least some positive impact on science outcomes. When a mother has a college degree it increases the odds that her daughter will have a science major in college. When a father has a college degree, it increases the odds that his daughter will expect a science degree by the age of 30. Similarly, mothers and fathers who have professional occupations are more likely to have daughters who are in science occupations or who plan to have one by the age of 30. A father's educational expectations and desire for his daughter to attend college after high school also increase plans for a science occupation and degree. Few have considered the influence of fathers in research on minority women in science and results attest to the importance influence that is working here. Finally, the study's results show that young African American women who discuss school courses with their parents are more likely to have a science major in college. Although a majority of the family effects in Table A.5.3 are positive, some are negative.

Community-Related Characteristics and Effects on Science

A number of variables measuring community-related variables (including volunteerism and religious participation) are shown for young African American and white women in Table A.5.4. Results suggest that African American and white women are similar on half of these characteristics. There are no race differences on items measuring, for example, non-school, civic (community), or church-related volunteer work. However, the young African American women are more likely then their counterparts to have friends who think it is important to do community work, to attend religious activities, and to have friends who think it is important to participate in religious activities. The young white women report higher scores on two of the variables—unpaid volunteer/community work, and volunteer work with youth organizations.

As with my earlier look at school and family characteristics, I find that each of these community-related characteristics is associated with some of the science outcomes for young African American women (as shown in Table A.5.5). Like the family variables, the community-related variables appear to be particularly cogent in their association with science achievement (as opposed to science access or attitudes).

The community-related characteristics that are most often associated with science access include, for example, importance (among friends) of doing community work, volunteering with youth organizations, receiving a community service award, and attending religious activities. Usually (but not always), those with more involvement in community-related activities score higher on the science access outcomes. For example, when the young women have been active in volunteer work with youth organizations they are almost twice as likely (27% vs. 14%) to expect a science degree by the age of 30. Over half of the young women who volunteered with a church group in the last twelve months were in advanced or accelerated science classes in 8th grade but less then one-third of those who did not do this volunteer work were in these classes. In a related way, those who attended religious activities more frequently were more likely to have greater access to science with, for example, frequent religious activity attendance doubling (26% vs. 13%) the likelihood of having coursework in chemistry in the 10th grade.

Each of the community variables, with the exception of volunteering with youth organizations, is associated with a good number of the science achievement outcomes. For example, performing unpaid volunteer or community service work is associated positively with all but one of the science achievement variables. Young women who are active here are more likely to get better science grades (in 8th and 10th grade) and score high on standardized exams (in 8th, 10th, and 12th grade). Some of these differences are dramatic. When women are doing unpaid volunteer or community work, for example, they are four to five times more likely to be in the top quartile on standardized science exams in 8th, 10th, and 12th grade. Having friends who think it is important to do community work is as likely to be associated with higher science achievement as it is to be associated with lower science achievement. But those who have friends who think this activity is important are over twice as likely to report a current (or most recent) occupation in science in the 2000 survey year. Note also that both of the religion-related items (attending religious activities and having friends who think it is important to participate in religious activities) are strongly related to reporting a science occupation in that year.

Although most of the community-related activities are associated with more positive science attitudes, this is not always the case. For example, volunteering with a civic or community organization in the last twelve months is related (in a negative way) to four of the seven measures of science attitudes. And, having friends who think it is important to do volunteer work is related to only one of the science attitude outcomes, and this relationship is negative. The results show that the important science outcome, planning on a science occupation at age 30, is positively associated with two of the community variables—volunteering with a church group in the last twelve months and being encouraged by someone else to do volunteer work. Finally, it is important to note the large number of community-related characteristics that positively relate to the young women's attitudes about looking forward to science class and working hard in science class.

Results from the multivariate logistic analysis examining effects of community variables on chances of young African American women scoring high on science access, achievement, and attitudes (as measured in the 12th grade and beyond) are presented in Table A.5.6. Each of the models explains a significant amount of variation in the science outcome (note the significant chi-squares). Two of the community-related variables (doing non-school sponsored volunteer work and attending religious activities) are seldom related to the science experiences but are negatively related in the cases where they are significant. There are, however, several strong and positive effects of valuing the importance of religious activity and participation. For example, valuing these activities increases chances of having a science major (in 1994) by a factor of 1.72. It increases chances of having a degree in science (in 2000) and having a science occupation (in 2000) by factors of 2.81 and 1.96, respectively. Other community variables are also important for positive science outcomes. These include, for example, performing unpaid community work in last two years, having friends who think it is important to do community work, and receiving a community service award. Many of these community/volunteer characteristics have positive consequences on more then one science outcome. Having friends who value community work is one of the most striking examples here. It positively influences three of the seven science outcomes (enrollment in 12th grade science classes, expectation of a science degree, and holding science occupation). One of the largest effects in the model (even larger then the effects of socio-economic status) is the effect of receiving a community service reward on respondents' standardized science scores. Thus, the NELS data support my earlier arguments about community work and community capital. Young African American women who engage in community and religious activities are often more likely to have positive science experiences.

As with the school variables, I examine the influence of a set of family and community variables in models that include the other factors influencing science experiences (school and peer). Results in Table A.4.4 show that the many of the family and community variables continue to have a positive and significant influence on young African American women's science outcomes, even when school and peer variables are taken into account.

Findings from Knowledge Networks

The analysis of Knowledge Networks data begins with a look at the quantitative measures from the web survey. Here we created a number of indicators that were similar to the NELS measures as well as a number of unique indicators asking, for example, about parents' degrees in science and closeness to parents. The examination of Knowledge Networks data also includes a look at the qualitative indicators of family influence on science experiences.

Quantitative Knowledge Networks Analysis

In a recent analysis (Hanson, 2006a), I used multivariate logistic and OLS (ordinary least square) regression models to examine the influence of the Knowledge Networks family variables on science experiences. Findings from this analysis are presented in Table A.5.7. Findings show that there are significant family influences for each science outcome. The most important family influence is "family encouragement in science." This family variable has a significant positive effect on each of the science outcomes. An examination of the standardized regression coefficients (not shown here) and effects on the odds suggest the family encouragement effect is invariably one of the largest effects in the model. For example, in the OLS models, standardized coefficients (not shown in table) show that an increase of one standard deviation in the family encouragement variable increases the "good in science" and "like science" attitudes held by young African American women by over two-tenths of a standard deviation. In the logistic models shown in Table A.5.7, each unit increase on the family encouragement variable increases the odds of expecting a job in science at age 30 by a factor of 3. A unit increase on this variable also increases the odds of hoping for a science occupation at age 30 by a factor of 2.5. Studies of social processes rarely show such consistently significant and substantive effects.

Several other family variables are significant in 50% or more of the equations. These include father has a degree in science, closeness to mother, closeness to father, importance of family in respondent's future, and importance of work in respondent's future. All but one of the family variables (mother has a job in science) is significant in at least one science outcome equation. It should be noted that variables measuring the influence of mothers and variables measuring the influence of fathers are both important in these science equations.

Overall, the nature of the family effects is as expected. Family variables such as family's encouragement in science, mother's degree in science, father's job in science, father's educational aspirations, closeness to mother, closeness to father and importance of work in one's future have significant and positive effects on young African American women's science experiences.

Some family variables worked in an unexpected manner. In the case of the science grades equation, mothers' educational aspirations and closeness to mother had a negative influence. Similarly, those who were closest to their fathers (or who had more involved fathers, or fathers with a degree in science) were not always the ones to do the best in science. Overall, however, the family influences on science outcomes were more likely to be positive than negative.

The two variables measuring the import of work and of family in the future present an interesting set of findings in these equations.

When the young African American women put value on family in the future, they are significantly less likely to think they are good in science, report that they like science, or hope for a science occupation at age 30. However, these attitudes involving value on family do not influence science grades, expectations for a science occupation, or the likelihood of feeling welcome in science. On the other hand, the attitudes about work variable has a significant and positive influence on four of the science outcomes—science grades, attitudes about liking science, reporting that one is good in science, and feeling welcome in science. Interestingly, the descriptive statistics (not shown here) reveal that the young African American women tend to be high on both family and work attitudes. And, they are significantly different than young white women on the work but not the family attitudes. Thus, simple conclusions about the importance of family and work attitudes for young African American women's science outcomes are not possible.

My Web survey was designed expressly for examining the influence of family and other characteristics on young African American women's science experiences. These data have fewer missing values and there is a larger sample than is available in the NELS data. The results show a strong and positive family influence on young African American women's science experiences. The evidence for the minority family being an important source of support for success in science is strong.

One of the questions included in the Knowledge Networks survey focusing on the experiences of young African American women in science asked explicitly about the importance of giving back to the community. The range of responses varied from "not at all important" to "very important." Results show that African American women are significantly more likely than are white women to give a positive response on this question. For example, 42% of the African American women (as compared to 24% of the white women) responded "very important" to the item.

Qualitative Knowledge Networks Findings

In a recent analyses (Hanson, 2006a), I also examined the qualitative data from Knowledge Networks. Here, the young African American women were asked a number of questions that examined the extent to which *they* saw a connection between family (or other) characteristics/experiences and positive experiences in science. A good number of the young women did not perceive influences (family or other) on their experiences in science. However, almost half of the young women gave some kind of response about family influence. Many of these were general. One young girl stated, "They let me try what I want to" and another replied, "My parents encouraged me to do good in all of the classes I took in high school." Slightly more than one-third of the women responded with specific examples of how their family encouraged (or failed to encourage) them in science.

A slight majority of the women *did not see influences of the family.* Many of the young women did not see other influences either. When asked whether and how their family influenced their experiences in science, a number of young African American women made statements such as, "I don't know," "They did nothing," "Can't remember," "No one said anything either way," "They don't say anything about it unless I ask them for their opinion. I like it that way," "I'm not a very [easily] influence[d] person," One girl noted, "Everyone's not the same and I may not have wanted to be in a science career but the other girls might. I didn't go into it [science] because of my own decisions, not nobody else's." Still another girl noted, "Because I have my own opinion and don't let anyone change the way that I feel." Another girl explained her interest in science this way, "I really don't know what it's from." One of the strongest statements of this type was from a young girl that liked science. When asked about whether (and how) her family influenced her in this area she reported, "No. Because I don't care about what other people have to say. It was my decision to be a nurse and it is my business whether or not I like science." These types of responses suggesting independence in thinking about school and academic interests were found in the white sample as well, although they were slightly less common. Ironically, it may be this very independence of thinking that allows the young African American women to go into an area where so many of them do not feel welcome. One girl even suggested that it is her parents who encourage this independence. She puts it this way, "My parents always told me to not worry about what other people say or do. Keep my head up and keep trying. Sometimes it's hard especially when it's your teachers that are supposed to be helping you." The last part of the statement attests to the fact that this young girl does not feel the support of her teachers. Others reflected on this same family pressure toward independence. For example, one girl alludes to independence and agency in response to a general question about influences, "I was always encouraged to do the things I wanted to do. I was told by my grandmother that I could be whatever I wanted to be if I committed myself to it and not lose my focus on the objective." On another more general question about influences, one young woman spoke of the influence of her family and others but also her own independence, "My teachers, counselors and family are able to witness my abilities in science. They encourage me but yet and still, my interest in what I want to do is really what counts."

At least one young woman began to see *connections between significant others' actions and her science experiences* as she filled out the survey. She reported, "In retrospect, the positive feed-back and encouragement must have planted a seed. I wasn't thinking of it as such then, but now I can recognize that a love relationship between me and science was developing then through positive reinforcement from influential people in my life." Another girl noted, "When someone shows interest and gives you positive feedback, naturally you will get interested." Finally, one of the young women was very explicit in saying, "Because the people around you shape and mold the kind of person you become." This way of thinking is an important one. If people see connections between family/significant others' actions and positive science outcomes then it is more likely that they themselves will be more active in encouraging science among other young minority women.

As noted above, a good number of women saw the *influence of family on their science experiences.* Although this was not a majority of the women, it was a significant number (over one-third) and their statements provide important insights into young African American women's perceptions of how their families influence their success in science. Examples of explanations here include: "Even when I was struggling they encouraged me and told me I would get it and overcome," "Because they encouraged me to explore the field more, by making it sound interesting," "Because I was always encouraged to do the things I wanted to do and was told by my grandmother that I could be whatever I wanted to be if I committed myself to and did not lose my focus on the objective." Some young women were very specific about what their family did. For example, in response to the question about how families influence them one girl wrote, "Just by helping me with my projects and by sending me to extra classes and programs in the summertime— one at Xavier College in New Orleans called SOAR and another at University of Alabama called MITE (Minority Introduction to Engineering)." Another girl was very specific about what her family did and saw its impact on her success. She reported, "At a young age my parents got me a chemistry set and kept children's encyclopedias around. I read them each several times. They also sent my sister to Space Camp and my other sister and I went to computer camps as young children. For my science projects, my father used to take me to UNC Charlotte library back in elementary school. One year I won 2nd place in the school science fair."

Mothers (and other women in the extended family network including grandmothers and aunts) were often noted. Thus although the quantitative analyses did show an influence of fathers, the young women seem to be more aware of their mother's (and other female family members) influence. Additionally, recall that a good number of the father's influences involved negative effects. Some examples of statements showing positive feedback and support from mothers and other female family members are:

> "My mother seems to have always known that I would become a physician. Consequently, she has always encouraged me in scientific endeavors. For example, she enrolled me in a Saturday science exploratory program at a place called the Math and Science Center for several years."

> "My mother and grandmother both were very interested in science-anatomy themselves and I would often listen to them talk about it."

> "Because a lot of my family is nurses."

> "My mom told me to do my best."

> "My family is full of women who either enjoyed or excelled in science."

> "My mother is an educator and she would do all she could to help me whether it was from what she knew or finding out from someone else in science if she didn't know. My mother has always encouraged me to do my best in whatever I did."

> "My mother never minded me using her kitchen utensils to dig up insects and worms to explore."

> "My aunt is always telling me that I am good at it and that she thinks I'll have a career in it because it comes easy to me. She encourages me to see things differently and try experiments at home."

> "My mother was and still is my biggest cheerleader. Her encouragement kept me going even though there weren't may other people to help or encourage" (in response to a general question about influences).

This young woman's mother was obviously one of her only sources of support in science. Some young women reported support from mothers but frustration in the difficult science curriculum. Recall that most research shows lower science grades on the part of young African American women relative to white women. One girl reported, "My mother and I would have conversations about things, how they worked, what they were made of. She would get really frustrated with me because she knew that my grades did little to reflect my intellect."

Some young women *saw family influence and saw the lack of external support* for young African American women in science. One woman reported: "I have a cousin that has a Ph.D. in science. She has [written] in journals and talked to me about the different stereotypes about African Americans in this field."

A few young women reported *positive influences associated with being an African American*. For example, one girl replied to the question about family influence with "[They] tell me about Chronology of Black History." Another girl reported, "My father had posters of African American women scientists up in my bedroom at his house. He took us on field trips to science museums and other cool places." But again, the young women often saw science as a less than welcoming place. The same girl who had posters of African American women scientists on her wall reported, "He (her father) actually tried really hard, but it was kind of him swimming against the tide of the world."

Some responses suggested that *families saw science as a way to get ahead*. For example, "My parents told me that science would always have employment," "My parents wanted me to make a better future by going into science and being a doctor," and "For my family it seemed that they were happy because it [science] would somehow bring prestige."

Finally, there was a small group of respondents that talked about *family influence in a negative way*. One young woman reported, "The negative feedback I got especially by my teachers and the women in my family were a great influence on my losing interest in science." Some of the responses suggest gendered images of science. For example, "My mother and grandmother upon reviewing my scores would tell me that it doesn't matter if I scored poorly in science since it is a "man-related" topic.

Other Qualitative Insights from Knowledge Networks

The young women in the Knowledge Networks survey were asked a number of questions about influences on science. In the discussion above, we looked at answers to a question that asked specifically about family influence. As we look at the answers to other, more general questions, we see that the family consistently comes up in the young women's discussions. In fact, one of my reports (Hanson, 2006b) showed that young women in the Knowledge Networks survey were more likely to report issues in their science experience when they thought of them on their own and were not asked directly. Interestingly, this appears to be the case with the questions on the family, as well. When the young women were asked general questions about science, science classrooms, and influences, their responses are peppered with statements about the family. For example, in response to a question about whether anyone else (in general) has encouraged them in science, one young girl writes, "mostly family" another writes "not outside of my family" and still another writes "No, my success came mainly from the support of my family."

Some comments contrast the negative school climate with the positive influence of the family. In one comment, a young girl suggests that teachers do not encourage ALL students (of all races and sexes) in science. She then adds "Parents are the key to students living up to their full potential … nobody else holds the greater responsibility then parents." As the literature above suggests, members of minority communities see and experience the bias in U.S. systems (including education) and turn to the family and community as their major source of support. One young girl writes, in response to a query about things that happened in the classroom that made her interested (or not interested) in science, "Nothing. I hated high school. A lot of my aunts were nurses and have retired in the last ten years. My grandmother was a nurse, so it's basically just in me to become one and carry on the tradition." This young girl's response not only speaks to the influence of family but of the family obligation to go into science. Others also alluded (with pride) to a family tradition. One girl writes, "School and learning is a very important part of my family's life. My mother was a teacher for more than 20 years. She is now a high school college counselor and advisor. My aunt (my mother's sister) is a grade school teacher and has been for over 20 years. My other aunt (also my mother's sister) is a Professor at the University of Minnesota. My father is an engineer who holds 2 graduate degrees. My mother also holds 2 graduate degrees." Others also spoke of the prestige that science brought to the family. In response to a general question about influence of teachers, students, and family, one young girl wrote, "For my family it seemed that they were happy because it would somehow bring prestige." But she adds, "They really were not interested in what it really involves or how I felt about it." Still others not only mention family influence but a broader tradition of blacks in science. One young woman admitted that she has felt the same way as the girl in the vignette (she doesn't belong in science) but she counters that with, "my mother has told me a lot about the great black scientists."

Some young women who had negative school experiences look back and wish that their family would have become involved. In her response to a question asking about examples of things that happened in the classroom that made the respondent interested (or uninterested) in science, one young girl wrote, "Terrible teacher who did not care whether I succeeded or not. Looking back, my mom should have intervened."

Responses to the question about why students liked or did not like science often included references to the family. One girl writes that, "my mother is into health and health issues—we would do various projects like melting ice, working with chemistry sets—to me that was fun spending time with my mother." But, as in the responses above that contrast the family experience with the school experience, she goes on to say that she disliked science in high school because of the teacher. Another young woman makes an analogy about science that compares it to a present from an aunt: "Science is like opening a present from your favorite aunt. You just can't wait to open it because you know that there is something wonderful and unique inside."

In the review of the literature and in my analyses of the quantitative data, I noted the disadvantage of lower SES groups in the science attainment process. Class, like race and gender, creates an obstacle to success in science. Some of the young women brought this issue up and realized the interconnectedness of class and family context. One young woman writes (in response to LaToya's negative experiences in the science classroom vignette), "She's right. Most black students don't feel they can accomplish much. It comes from various things happening in the home—black children aren't taught to really try hard things—there can be a shortage of money." Others also saw a lack of support in the black family and connected this to race more than gender. In response to a question about whether African Americans or women had a harder time in science, one young woman wrote, "African Americans, because a lot of African American children don't get the support from their parents like they should." Another young woman asks that teachers be more concerned since sometimes parents are not. She wonders, "If teachers can be a little more concerned with the students' education because sometimes the parents don't do their jobs at home so we need the teachers for that encouragement." One young woman suggests cooperation between the two in her words, "The teachers and parents are the ones who can encourage someone in science."

Finally, as suggested in the chapter on schools, not all of the young African American women had negative school experiences. Some who had positive school experiences connect these to the family as well. In response to a general question about influences, one young woman writes, "my interest started at home and the school gave me more insight in the field of science." Others also saw their educational interest in science start with the family, sometimes with the everyday context of life in their home. One young girl's comment (when asked whether there was a time in the educational process that young women become interested in science) was, "when you reach junior high school you become more aware of your surroundings and your curiosity might begin to sink in. I know mine did. We had all kinds of pets. My dad really loves animals and plants. We had snakes, rabbits, birds, turtles, etc. You name it. We had trees, plants covering the yard and house and porch. The kids in the neighborhood called ours the jungle house. So I had an interest at an early age of these things and how they came to be."

Conclusions

My research has shown the considerable interest and involvement of young African American women in the science domain, in spite of the fact that teachers and schools often do not see them as science talent. In this chapter, I have suggested that the African American family and community might be an important source of agency in this realm. In spite of negative stereotypes about African American families (that are often shared by schools), my past work (Hanson and Palmer-Johnson, 2000; Hanson, 2006a) and that of others (Luster and McAdoo, 1995; Foster and Perry, 1982; Mau et al., 1995; Higginbotham and Weber, 1992), suggests African American families are often a source of strength and resilience in the educational realm (especially for young women). African American communities, in general, can be seen as a source of community capital given their focus on community and their investment in children, especially females (in contrast to the individual focus and greater investment in males typical of white communities). Community investment in children, I argue, results in young people's desire (and obligation) to give back to the community through achievement in elite areas such as science. Success is seen as a group effort that involves support of family and community. Additionally, I argue that gendered aspects of African American communities involving the integration of women's work and economic roles, and the focus on education and occupation as the major source of mobility for young women are important elements for young African American women's success in science.

My research findings provide considerable evidence for the African American family and community as a mechanism for support in science among young African American women. Data from the U.S. Department of Education's NELS surveys show that young white women may have more socioeconomic resources than African American women, but the reverse is often the case for family variables measuring parent's educational expectations and family involvement. Additionally, although there are no race differences on some community variables, others such as peer value systems involving community work and religious activities as well as respondent's involvement in religious activities are reported with more frequency among the young African American women (relative to the white women). Many of these family and community characteristics are shown to be related to success in science. My hypotheses about focus on community and giving back were supported by data from the Knowledge Networks Web survey that showed young African American women in the sample were more likely than young white women to feel that they needed to give back to the community.

Findings from the Web survey conducted expressly for this research on minority women in science provide even more support for the input of family. Quantitative analyses show the import of family encouragement and closeness for the young women's success in science. Qualitative analyses show that the young women often feel very independent in their pursuit of science. This is not surprising given that these are young adolescent women and given the unique African American gender system stressing independence among young women. However, significant numbers of young African American women make it clear that family influences are strong. Interestingly, reflecting on these issues in the survey made some young women see the influence for the first time. Many of the young women saw considerable encouragement, especially from mothers, aunts, and female relatives. A good number of the women saw this support as being critical given the absence of external support. Some even saw positive influences associated with being African American. Interestingly, it was when the young women were asked about school experiences in science that positive comments about family support often appeared. Young women often reported that it was mostly family, and few outside of family, that supported them in science.

The unique contribution of this research comes from the Web survey designed expressly for studying African American women's science experiences. Perhaps the most important data collected is the qualitative data that allows the young women to report their thoughts and experiences in their own words. Here we can read in report after report about mothers who were the young woman's biggest cheerleader, families full of women who were interested in science, fathers who hung up posters of African American scientists, and mothers who let them use kitchen utensils to dig up worms. In spite of this support, many of the young women felt discouraged in science. As one young woman reported, "He [my father] actually tried very hard [to support my science interests], but it was kind of him swimming against the tide of the world."

6 Influences—Peers

"I Know Plenty of Girls at My School [Who] Love Science."

There is only a limited amount of literature available on peer effects for young African American women in the science domain. In the discussion below, I briefly review the literature on the importance of peers for education (and science education) achievement. I then review literature on gender, race, and peer experiences in education (and science education). This literature (together with a consideration of oppositional and other theories on race, sex, and peer processes in education systems) helps us understand general peer effects and the peer culture that African American women experience. It also sheds insight on the ways in which African American female peer culture is distinct from African American male peer culture and white female peer culture. Finally, I examine the limited literature examining peer influences on African American women's education (and science education) experiences.

It should be noted that the term "peer" is used in a very broad way in the literature. The research examined here considers a number of influences including those traditionally thought of as peer effects. These include the impact of social networks (social capital) and friends located in neighborhoods, schools, and classrooms. Literature that considers influences of other students in the classroom, school, and neighborhood that are not necessarily friends or members of the young person's immediate social network is also examined.

Background: The Literature on Peers, Schools, Gender, Race, and Science

Importance of Peers for Education Outcomes

Research on adolescence suggests that peer relationships gain import as youth move from childhood to adolescence and the influence of peers is one of the strongest influences during this period (Furman and Buhrmester, 1992; Goodenow and Grady, 1993; Brown, 1990). In fact, youth spend more time with peers than with adults (Brown, 1990). It is not surprising, then, that young people cite their friends as their strongest source of support (Furman and Buhrmester, 1992). Daily relationships with peers in schools are important factors in teen's perceptions of social reality and personal identity (Kinney, 1993). These factors together with the time spent with friends, and the value placed on their support, are indicative of the potential for peer influences in the educational realm (Stake and Nickens, 2005).

Research on educational attainment has consistently shown an influence of peers on young people's educational values and outcomes. Early research examining school influences (Walberg, 1984) and the status attainment model (Alexander and Campbell, 1964; Campbell and Alexander, 1965; Haller and Butterworth, 1960; Sewell and Hauser, 1972; Hanson and Ginsburg, 1988) showed the import of peers for educational attainment. Later research also confirms the importance of peers for educational outcomes involving attitudes, school adaptation, and academic competence (Fredricks and Eccles, 2006; Corsaro and Eder, 1990; Levitt et al., 1999), as well as general educational motivation and outcomes (Berndt et al., 1990; Gustafson et al., 1992; Ide et al., 1981). Peers have also been shown to influence academic achievement motivation, school satisfaction, expectations for success, and academic performance (Goodenow and Grady, 1993; Jacobs et al., 1998; Manis et al., 1989; Ryan, 2000).

Hanson (1994) found peer influences affect lost talent (reduced educational expectations and achievement) and this influence varies by race and sex. Research on the academic achievement of African American youth has shown the importance of peer influences (Clark et al., 2003). In fact, there is research that suggests that peer influence is particularly important here (Levitt et al., 1999). Some have argued that the social capital aspect of peer groups works to the disadvantage of black youth in the educational context (Sewell, 2000; Sampson et al.,

2002). In a related way, literature looking at neighborhood effects provides evidence of the influence of neighborhood peer groups via social
capital on a wide variety of outcomes (including education) among
adolescents. Factors involving social control and resources also explain
these neighborhood peer effects (Sampson et al., 2002) and the potential disadvantages of minority youth on these resources.

Race is not the only status influencing peer groups in the education
system. Sex is also a master status and, increasingly, it is young women
who are advantaged in the educational context. Research suggests that
today's young women are increasingly outperforming young men on
high school and college grades, attendance, and degrees and these differences are greater in minority populations (Lopez, 2003; Cho, 2007;
Peter and Horn, 2005). Sex (male/female) is a major source of diversity
in African American peer groups. Research on educational and occupational achievement in the African American community suggests that
young African men and women have different expectations and experiences with racism and resistance. These experiences filter into peer
cultures and influences. The greater expectations for young African
American women as well as their greater chance of success in educational and occupational spheres (relative to young African American
men) are well documented (Higginbotham and Weber, 1992; *Journal
of Blacks in Higher Education*, 2001; U.S. Bureau of the Census, 1995).
These trends go along with reports of less discrimination for young
African American women than men and less pressure for young African American women (than men) to "act white" (*Journal of Black in
Higher Education*, 2001).

Importance of Peers for Science Education Outcomes

Similarly, early research using the status attainment approach found
that peers influence science-related outcomes (Hanson and Ginsburg,
1988). More recent research also shows this influence (Lee et al., 2003;
Rani, 2000; Hanson, 1996). Peer pressures and feedback are particularly relevant given my focus on science education in the high school
years. It is during this period that the young person is forming their
identity and feedback from peers is one of the strongest influences for
adolescent youth. It is this identity (so heavily influenced by the attitudes and feedback of peers) that will be an important element in predicting success or failure in school (Sadowski, 2003).

A wide variety of science-related outcomes are influenced by
friend's attitudes toward science. These include, for example, enjoy-

ment of science, science activities, attitudes about a career in science, and science course-taking (Simpson and Oliver, 1990; Jacobs et al., 1998; Kelly, 1988). Researchers have found the correlation between friends' attitudes about science and own attitudes about science to increase throughout adolescence (Talton and Simpson, 1985).

Peer cultures are particularly important for young people's science interests and experiences. Given, the import of these cultures for identity, young people wish to avoid being labeled as members of unpopular groups (Kinney, 1993). Unfortunately, there are a considerable number of negative stereotypes about science. Not only is science seen as being for old white males but it is also perceived as being boring, and those with an interest in science are sometimes labeled as geeks and nerds (Betz, 1997; Margolis et al., 2003; Jeffe, 1995; Harlen, 1985; Vinchez-Gonzal and Palacios, 2006; Constant, 1989; McDuffie, 2001; Palmer, 1997; Powell and Garcia, 1988; Kinney, 1993). LaFollette's (1988) study found a considerable number of negative images and stereotypes involving women scientists. Given this often negative image of science and the desire to avoid being labeled as a member of an unpopular group, this presents a problem for many young people who seek peer support but are interested in science. Some minority group's advantages (e.g., Asian American youth) in science have been argued to come from peer support and help networks (Kao and Thompson, 2003). For other minority youth, it may work in the opposite direction. Given the underrepresentation of African Americans in science, some have observed a "stereotype threat" surrounding science for African American youth (Maholmes, 2001). Stereotypes that depict scientists as white have been argued to be an important factor in keeping minorities out of science (Edwards, 1999). Solutions for creating interest in science often use peer approaches (e.g., creating college residence halls with similar peers) (Palmer, 1997). Some researchers examining science models separately for whites and blacks have found peer effects were limited to whites (Hanson and Ginsburg, 1988). However, others have found effects for African Americans as well (Fries-Britt, 1998).

Fries-Britt (1998) found that high-achieving black students in a merit-based program for students in math, science, and engineering were isolated from other minority students (outside of the program). However, their membership in the program helped them establish peer networks among other high-achieving black students that helped diminish this sense of isolation and promoted their success in the program. The young minority students found a community of people like themselves with whom they could relate and seek support. Having this "critical mass" of other minority students who were interested in (and good in) science allowed them to blend in. This is often not possible in a typical science classroom where the pressure to fit in works against and not for success in science.

African American youth and their peers often view science as only available to the exceptionally talented (Pearson, 1982). These notions work to limit their interest and engagement in science. Additionally, it is important to recall the positive attitudes of African American youth toward math and science (see, e.g., Davis et al., 1989; Anderson, 1988). Thus, at least at early ages, there is not an anti-science culture among minority youth. At early ages, it is not African American peer cultures but rather the lack of opportunities, resources, role models, and support networks that work to discourage young African Americans in science (Pearson and Bechtel, 1989). Pearson's (1982) research on the origins of African American scientists suggests that African American youth surrounded by African American peers are not discouraged from science. In fact, most African American scientists came from predominantly black neighborhoods, high schools, and colleges. Pearson's suggestions for understanding the flow of African American youth out of the science pipeline involve factors such as standardized exams, role models, and teachers. They do not include peer groups with negative attitudes about science.

Gender is also an important factor in affecting peer influences on science outcomes and there is considerable research on the topic. In general, research (as noted above) has shown that young women (especially during the high school years) have higher commitment to and achievement in school than do young men. Hanson (1996) found young women are also more likely to have peer groups that put an emphasis on education and these peer groups are associated with success in science. However, the male culture of science is mirrored in the peer cultures of young men and women. Young women are exposed to peer groups that think of science as a male domain. It is seen as a domain that would not allow them to be feminine in traditional ways and that would compete with feminine ideals of marriage and family. These attitudes come to children via families and school systems (Hanson, 1996; Fox and Tobin, 1980; Sherman and Fennema, 1978). The hidden curriculum is a major factor in creating ideas about competence in science among young women (Hanson, 1996; AAUW, 1992; Oakes, 1990) and these ideas are powerful influences on young women's attitudes and decisions in the realm of science. These are relevant to the discussion of peers since the hidden curriculum helps create gendered peer cultures

Although peer support influences both boys' and girls' interest in science, level of support for science has been found to be lower among girls. Women's lesser interest in science has been connected to the lower levels of support among peers (Stake and Nickens, 2005; Jacobs et al., 1998; Rayman and Brett, 1993). Young women perceive less support for science among their friends than young men do (Kelly, 1988). They talk less about science with their friends and engage in less science activities with their friends than do young men (Kahle and Lakes, 1983; Jovanovic and King, 1998; Kahle and Lakes, 1983). Baker and Leary (1995) found that at a young age (8th grade) most young women like science but think their friends would disapprove of a science career. By 11th grade, the study found that while many young women did not perceive disapproval of a science career among their friends, they no longer liked science. Negative response from peers makes it difficult for young women to take classes that deviate from what is expected for their gender (Beal, 1994). Some have suggested that it is peer effects that are a major factor in reducing girls' interest in science (Rayman and Brett, 1995) and eventually college women's interest in science (Holland and Eisenhart, 1991).

Male peers also provide feedback that influences young women in science. Young women sometimes report that male peers do not make them feel welcome in science classes and activities. Surveys of young men show a tendency to see science as a male domain that is not appropriate for women (Erb and Smith, 1984; Stake, 2003; Terry and Baird, 1997; Farengar and Joyce, 1999; Greenfield, 1996).

When girls are interested in science, peer support has been shown to be a critical factor (Fish, 1979). Having a close friend in science can have a positive impact on young women's attitudes toward science and science careers (Leary and Baker, 1995). Stake and Nickens (2005) found that positive peer relationships (in the science setting and out) are critical in allowing both male and female youth to develop a perception of themselves as scientists.

Studies have shown the importance of support from female friends for young women's continuation in science (Boswell, 1985; Fox et al., 1979). An ethnographic study of college women showed that the peer group was a major factor in shifting young women's goals away from careers and interests in science toward romance and heterosexual relationships (Holland and Eisenhart, 1991).

Peer group norms are at least part of the explanation for the success of single-sex girls schools in recruiting young women into math and science courses, and, in general, in preparing them for nontraditional occupations (Fox et al., 1985; Schwager, 1987; Tidball, 1980; Rice and Hemmings, 1988; Lyall, 1987). A disproportionate number of women scientists have spent time in single-sex colleges (Tidball, 1975, 1986). The presence of a critical mass of women has been suggested to be an important ingredient for this success (Rayman and Brett, 1993). Additionally, programs that have focused on peer relationships and support have been successful in creating and maintaining interest in science among young women (Stake and Mares, 2001).

Gender, Race, and Peer Processes in the U.S. Education (and Science Education) System

In Chapter 4, on school systems and science experiences, I noted that education systems in the United States (and elsewhere) tend to reproduce the race and sex inequalities in the larger culture. As Bourdieu (1973) argues, this involves an arbitrary reproduction of culture through the process of social selection of students (by race, class, sex, etc.). Tracking and other selection systems that involve differential learning opportunities work to label and process students into groups that will have different chances of success by virtue of their placement in this selection system (Cooper, 1996; Oakes and Lipton, 1992; Kao and Thompson, 2003). Researchers have shown that certain courses (e.g., computer science) become labeled and experienced as white (or Asian) and male spaces (Margolis et al., 2003). What is important for us in this chapter on peer influences is the cultural climate surrounding students that are not expected to succeed, and who are grouped together. It is in this way that schools and the peer cultures created through tracking and selection sometimes work as obstacles, not aids to achievement (Johnson, 1997).

Race and sex are major categories in the educational placement system. It is within this context that Ogbu (1974, 1978, 1991) and others have observed minority peer groups that understand their poor chances of success in a white system and their eventual rejection (and peer pressure to cooperate in rejection) of that system. Race stratified education systems contribute to distinct "white" and "black" cultures. It is problematic, according to Fordham and Ogbu (1986) when "acting black" and "acting white" become identified in opposition to one another since "acting white" includes doing well in school. "Acting black" then becomes the opposite of that and thus implies not doing well in school. For some, this peer culture provides the prism through which they experience education.

Similarly, researchers (Fordham, 1993; Evans, 1988) have found young African American women resisting white female gender systems and the negative self-evaluations that come from white standards. Their resistance involves the creation of their own gender attributes involving male-type characteristics (as in "those loud black girls") and (as in Ogbu's 1974, 1978, 1991 research) less success in the educational system.

Some researchers have not found support for Ogbu's (1974, 1978, 1991) oppositional culture hypothesis. Instead, Ainsworth-Darnell and Downey (1998) found pro-school attitudes among African American students (at higher levels than whites) and, importantly, no negative peer effects. Mickelson (1990) also found pro-education attitudes among minority students. She concludes that minority youth do get frustrated with an educational system that does not work well for them and does not appear to bring returns. This frustration is reflected in their concrete attitudes about education. However, at an abstract level, they continue to place a high value on education as an ideal. These findings are not inconsistent with Ogbu's (1974, 1978, 1991) work given that Ogbu's primary focus is on concrete school behaviors and limited efforts due to perceptions of limited opportunity for success.

Although there is some disagreement on the oppositional culture hypothesis, research showing the impact of race biased educational systems on young minority student's achievement and attitudes is extensive (Cooper, 1996; Freeman, 1997). It is important to note, however, that there is considerable diversity among African American youth. McNamara Horvat and Lewis (2003) found diversity in African American peer groups and concluded that young African American students were able to find peer groups that promoted academic success. Kao and Thompson (2003) also found evidence for diverse peer groups available to African American youth and the ability to manage these peer associations in an effort to achieve high educational aspirations.

There is disagreement as to whether it is peer groups or close-knit friendship groups that are most important for minority youths' educational aspirations and achievement. Evidence exists for both (Wilson-Sadbury, 1991). The research suggests that peer effects are complex, especially when considering race and sex variation. For example, Bowman-Damico and Sparks (1986) found that young black women are primarily influenced by black peers, whereas other research suggests the import of white peers for minority youth (Wilson-Sadbury, 1991). What is clear is that the lack of racial diversity among teachers and in course content promotes the need for an alternate identity. This can create a feeling of non-inclusion among minority youth and thus peer-group resistance to the white system. If teaching staff and course content were more diverse, these alternate cultures would not be necessary.

This sense of an alien environment in the school and classroom and the race discrimination and race stereotypes that was discussed earlier result in strained relations between African American students and whites, including white students. White students follow the lead of school systems and teachers who view minority youth as having a lower learning capacity (especially in the advanced areas of math and science) (Anderson, 1988).

Work on science achievement provides similar conclusions regarding the impact of tracking and selection. Research suggests that this selection process in math and science begins in the early middle school years and has an influence on course taking and success in math and science in the high school years. Given the race and sex biases in these selection processes, the selection system creates stratified social groups that have an influence on young people's math and science attitudes and achievement (Eccles, 1997).

Research on tokens adds to insights about race, sex, and peer processes in the science education system (Kantner, 1977). When women and racial minorities enter into the white male science domain, they are often the numerical minority. This can lead to isolation and to tokenism by teachers and students. Tokenism is alienating and works to discourage women, racial minorities, and minority women in the sciences (Betz, 1997). Grandy (1998) found that persistence in science among capable minorities was negatively influenced by discrimination and the isolation from peers and campus organizations.

In spite of disagreement on process, researchers agree that minority youth are not provided the same skills, habits, and styles that white youth receive and which teachers reward (Ainsworth-Darnell and Downey, 1998) and that this is most likely due to structural effects in schools and neighborhoods. The notion of social capital is relevant to these ideas of power, selection, and grouping of youth in the U.S. education system. Placement in the system puts youth in networks with other, similar youth. Those placed high in the system have the advantage of positive social capital and networks that work to their advantage. Those placed low in the system, have the disadvantage of less social capital and networks that might work to their advantage. It should come as no surprise then, that researchers have found an advantage in science for African Americans in HBCUs (Project Kaleidoscope, 1991).

It is important to consider the potential agency of minorities and women in education (and science education) systems. As noted earlier, these minorities do not always accept all of the messages that they are given. Sometimes cultures of resistance and strength are formed among these young people in their peer groups. Additionally, the flow of certain resources in families and communities sometimes works to the advantage of females, racial minorities, and female minorities.

Hanson's (1996) study of the development of talent in science found that young women have the advantage on peer resources. They are more likely than young men to have friends who get good grades, are interested in school, attend classes regularly, plan to go to college, and think well of students with good grades. These peer resources are important factors in predicting which young women with interest and talent in science remain in the science education pipeline. Race and sex work together when considering peer influences on science education outcomes. Hanson (1996) found that African American women, on the whole, are more likely than other women to have school peer groups who put emphasis on school. For example in the senior year of high school, young African American women are more likely than other women to have friends who: are interested in school, attend classes regularly, and plan to go to college. As suggested above, these peer influences affect progress and interest in science.

A Consideration of Gender and Race: African American Women and the Peer Context in the U.S. Education System

As noted earlier, young African American women are not immune from the tracking, selection, racism, sexism, and tokenism that young African American men experience. These school contexts affect their peer culture, to some extent, in the same way that they affect peer cultures among all minority youth. African American women are tokens and outsiders, especially in higher education (Moses, 1989).

A critical item for consideration is the fact that young African American women have distinctive gender values relative to young white women. Many of these values, especially those related to work and the family, are associated with interest, success, and agency in science. African American women have a unique position in the African American community. In Chapter 5, I described a culture where African American

families put more emphasis on the educational mobility (and investments into education) of their daughters than their sons and this
emphasis on education for daughters is greater in African American
than white families (Higginbotham and Weber, 1992). Here, differences between men and women on educational achievement, rate of
labor force participation, and earnings are less than in the white community and in some areas (e.g., high school and college degree rates)
women are surpassing men (at a higher rate than in the white community) (Hanson and Palmer-Johnson, 2000; Lopez, 2003).

Thus, the peer culture for young African American women is not
the same as for young African American men or for young white
women. In fact, some have shown that these young women have more
androgynous gender roles and greater self-esteem, independence, and
assertiveness as well as high educational and occupational expectations
relative to other women (Andersen, 1997; Hanson and Palmer-Johnson,
2000; Hill and Sprague, 1999; National Center for Education Statistics,
2000a). All of these things are not only related to educational success,
but also success in science (Hanson, 1996). Strong personalities among
peers socialized in communities where women have always worked and
have often combined work with family may work as distinct cultural
capital for young African American women.

Some research has suggested that African American women exhibit
less fear of success (Murray and Mednick, 1977) and are less subject
to gender stereotyping and fear of rejection by men when considering
a career since minority men are more supportive of working wives than
are other men (Axelson, 1970; Scanzoni, 1975). They also have more
liberal sex-role attitudes. Research has shown that young African American women do not experience the loss in self esteem that other women
experience as they progress through the high school years, and this is
partially related to their androgynous characteristics (Buckley and
Carter, 2005). They date less and expect to marry and have children at
a later age. Finally, they express a greater interest in science than do
young white women (Hanson, 1996). All of these individual characteristics are associated with success in science (Hanson, 1996). In Chapter
3, I noted the positive attitudes toward math and science (and careers)
among African American women and their high representation in science occupations (relative to African American men and white women
(Leggon, 2003). Baker (1987) and others (MacCorquodale, 1984;
Murrel et al., 1991; Rubenfeld and Gilroy, 1991) found that the more
stereotypical male characteristics young women perceive themselves as
having, the more likely they are to chose a nontraditional career.

Studies of African American professional women show a similar climate of success in spite of race and sex statuses. Instead of double jeopardy, these women sometimes avoid discrimination because they are not seen as competition by white males and, to some extent, are not seen as female (Fuchs Epstein, 1973). The culture of professionalism, success, self-confidence, orientation to work, and non-traditional gender systems among young African American women is distinct from the white female culture and the African American male culture. It is within this context that young African American women make choices. They are influenced by surrounding networks of African American women with similar value systems.

Peer Influence in the Sciences: African American Women

These general characteristics of African American women and their peers may be part of the explanation for the fact that young African American women take more advanced math courses than their male counterparts (Matthews, 1984). They also have attitudes about math that are similar to those of white males and more positive attitudes about math as well as better science grades and more science participation relative to their African American male peers (Marrett, 1981; Matthews, 1980; Hart and Stanic, 1989; Clewell and Anderson, 1991). Any evaluation of peer effects in science, then, must take into account the characteristics of the young African American women who will act as influences. It is too simple to talk about a generalized oppositional culture within African American youth cohorts without a consideration of gender. Certainly, some of this culture may hold for young women, but the evidence seems to suggest that these young women form a community whose educational endeavors are strongly supported by the broader African American family, community, and gender system. The evidence suggests that they have positive orientations toward education and toward science in particular.

Colleges that have been the most successful in encouraging minority women scientists are women's colleges and historically black colleges and universities (HBCUs), especially women's HBCUs. Some of the arguments for this success have to do with diverse faculties. Importantly, these universities place the young women into a group of students and peers who are similar on race, or race and sex statuses. This peer environment is part of why these universities and colleges enhance self-confidence and success in science (Jordan, 1999; Pearson et al., 1999). Rayman and Brett (1993) found that young African American women at Howard University (an HBCU) rated themselves higher on mathematical ability and self-confidence than did young African American women at Wellesley (a single-sex college where they were the racial minority). The young African American women at Wellesley had few friends to turn to and often felt excluded in the classroom. Results suggest that the young African American women perceive race as a larger barrier than gender and would rather be with peers who are similar on race than with those who are similar on gender.

Additionally, the young African American women in science classes at Howard University were not intimidated by men in the classroom (Rayman and Brett, 1993). The researchers speculated that this may be due to the proportionately larger number of women (than men) in these classrooms (they constitute a critical mass) or the unique cultural relationship between African American women and men.

Recall, also, that African American women tend to have positive attitudes about science (often more so than their male or white counterparts) again, suggesting a positive source of peer influence (Hanson, 1996). Thus female, African American peers systems might be a source of agency, not failure in the education (and science education) system.

Gender and race differences in science are a function of social structure and context (Catsambis, 1995). The context of cohorts and peers with strong personalities, little fear of science, and emphasis on education is one that might be associated with success for African American girls in the education (and science education) systems. Both the structural impediments (involving race and sex) discussed above and the resources and agency discussed here are important elements of young African American's peer systems and science experiences.

In sum, the literature on peer influences in the education (and science education) system presents a complex set of processes. Peers are important influences, especially during the high school years. Their impact on attitudes about and interest in science is considerable. This influence varies by race and sex and commonly held stereotypes about the nature of race and sex effects do not always hold. Assumptions about negative African American peer influence, minority youth with oppositional cultures, and low social (and cultural) capital on the part of women and minorities are unable to explain the high value placed on education by many minority youth and (more specifically) the high level of interest in science among young African American women. Tracking and selection processes in the U.S. educational system along with racism and sexism do set up barriers to women, minorities, and minority women. These youth do sometimes manage, however, to build peer cultures and have peer influences that provide agency and support.

The research on race, gender, peers, and education (including science education) suggests an important, but complex role of peer groups in educational processes. However, much of the research ignores these complexities and does not consider how peer influences might vary across race and gender subgroups. The focus is often on general student groups, or males, or females, or whites, or blacks. I now turn to an analysis of the secondary NELS data and the survey (Knowledge Networks) data to examine the role of peers in young African American women's science experiences.

Findings from NELS

The analysis of peer effects begins with an examination of means and standard deviations for a number of peer variables. The sample includes African American (and white) females who were high school students in the NELS sample.

In Table A.6.1, means are presented for nine peer variables. These variables measure the importance of having strong friendships but also the characteristics of friendship groups. Do the young woman's friends think good grades are important? Do they find it important to study, to be popular, to have sexual relations? How important are friends in the science decision, do parents know the student's friends, and do these friends hope that the respondent will attend college after high school? There are significant differences between the African American and white female students on all but two of these peer measures. There is no race difference on importance of friends in the science decision. Approximately 20% of young African American and young white women say that their friends are very important to them when it comes to making decisions about science. Similarly, the two groups of young women are equally likely to have parents who know their closest friends' parents.

Results show there are race differences on peer characteristics, and it is often African American women who have the potential advantage. They are less likely than are their white counterparts to think it is important to have strong friendships. (It is unclear whether this is an advantage or disadvantage in the science education process.) They are also less likely to have peers who think it is important to be popular (popularity often has a negative effect on the science decision) or to have sexual relations. Additionally, the young African American women are *more* likely to have friends who think it is important to study, who get good grades, and who hope that the respondent will attend college after high school. The white females have an advantage on only *one* of the peer items. They are more likely to have friends who think that it is important to finish high school. The race difference here is not large, however, with a majority (89% for whites and 82% for African Americans) of both African American women and white women having friends who think it is important to finish high school.

These results are interesting in that they conflict with the popular notion of negative peer influences in African American communities and the anti-white, anti-academic culture that has been described for African American youth. The results are similar to those found in the examination of family literature and family effects in the NELS and Knowledge Networks data. My findings do not support negative stereotypes about young African American women's families or their peer groups. Literature supporting these stereotypes often relies on limited empirical evidence. Additionally, researchers tend to focus on male youth or combined samples of African American male and female youth without regard for the complex ways in which race and gender work together. Finally, few researchers have focused specifically on the science outcomes that are examined here. Rather, they have focused on broader academic outcomes.

Figures in Table A.6.2 show the science experiences of young African American women with various peer characteristics. These data help us describe peer arrangements for those who do well in science. We see here that the two most important peer characteristics in describing the African American women who are engaged in science have to do with having friends who think it is important to get good grades and having friends who think it is important to finish high school. In most (but not all cases) those young women who have these types of friends are more engaged in science. Having friends who think it is important to finish high school is the most important peer factor in these analyses. For example, almost twice as many young African American women with friends who think this is important were in advanced science courses in the 8th grade (40% vs. 23%). The two peer variables that are the least important in describing the African American women who were engaged in science involve importance of friends in the science decision and importance of sexual relations among friends. Peer arrangements appear to be most important in describing differences amongst young African American women on the science course taking (access) and attitude variables and less important in describing differences on the achievement variables. Interestingly, some variables that have been shown to be related to success in education for combined samples (or white samples) have the reverse effect here. For example, in most cases, the young African American women who do well in science are less likely then other women to say that their parents know their closest friends' parents. This suggests parental involvement works in different ways in the African American community. There may be some complex relationship between peer and parent characteristics. For example, young people who are having trouble in school or in the community may be likely to have parents who know their friends' parents.

Findings on peer characteristics of those who are engaged in science support the notion that peer groups influence science experiences for young African American women. Two peer factors stand out as being the most important here, having friends how think it is important to finish high school and having friends who think it is important to get good grades. In my examination of peer variables above, I showed that young African American women actually have the advantage over young white women on the grades outcome and are only slightly less likely to have friends who place value on finishing high school. (It should be noted that a large majority of the young African American women's friends [80%] stress this.)

I next examine the peer effects in a multivariate context. Results in Table A.6.3 show the effects of each of the peer variables when other peer variables, along with SES, are controlled. As in other logistic analyses, I examine the effects of the peer variables on the odds of reporting "yes" on a number of science outcomes. As in the descriptive table, having parents who know one's closest friends is associated with less engagement in science. Stress on popularity also continues to take away (in most cases) from science success. The direction of some variables that had positive associations in the descriptive data changes when other peer variables and SES are controlled (e.g., some of the effects of having friends who stress finishing high school). When other peer variables and SES are controlled, the peer variable that has the strongest positive impact is the variable measuring the importance of having strong friendships. Young women who agree with this more then double their odds of getting a postsecondary degree in science and of planning on a science occupation at age 30. Thus, the friendship variable that might have worked against science engagement (if peer influence was a negative factor) works for positive science outcomes suggesting a positive impact of strong friendship groups for science experiences.

As with the school, family, and community variables, I examine the influence of a set of peer variables in models that include the other factors influencing science experiences (school, family, and community). Results in Table A.4.4 show that the peer variables continue to have a positive and significant influence on young African American women's science outcomes, even when school, family, and community factors are taken into account. As noted earlier, being popular takes away from positive science experiences. Effects of placing importance on strong friendships continue to be significant and positive when other causal factors are included in the model. Interestingly, importance of popularity (not placing value on friendships) appears here as a powerful factor that detracts from science experiences.

Knowledge Networks

I turn now to an examination of the peer variables and their effects on science in the Knowledge Networks Survey, beginning with the quantitative analysis.

Quantitative

An examination of Table A.4.5 shows the effect of one of the peer variables in the quantitative Multiple Classification Analysis using the Knowledge Networks data. The peer variable measures whether students at your school tend to encourage you in science. This is a multivariate analysis so effects of family, community, school, and SES variables are controlled. The peer variable works in opposite ways on science occupation variables and attitudes about science variables. Having students at school encourage you in science contributes to lower expectations and hopes for a science occupation but has a positive influence on feeling good in science and feeling welcome in science.

Qualitative

Next, I examine open-ended answers to questions in the Knowledge Network survey. The questions focus on encouragement in science from students in high school as well as factors that make for positive science experience. There are five broad themes in the young women's responses involving peers and science. These include:

- Friends and schoolmates are not interested in science and do not provide support or encouragement in science;
- Friends are interested in science, talk about it, and provide support and encouragement in science;
- Race and peer groups work to influence science;
- Gender and peer groups work to influence science; and
- Peer cultures view science as dumb and something that is for geeks and nerds.

In the pages that follow, I describe each of these themes and provide some quotes from the young women that help exemplify their feelings about peer influences on science.

Friends and schoolmates are not interested in science and do not provide support or encouragement in science.

The first theme in the open-ended Knowledge Network data on peers focuses on the fact that the young women's friends and schoolmates are not interested in science and do not support or encourage them in science. Some of the young women note a *general lack of interest in school among their friends*:

> "Students in my school weren't really about school. They were about getting their work in and passing but not so much as to actually be good at a subject and furthering it."
> "The kids at my school play too much, so we don't get much done in class."
> "We usually don't talk about school a lot. Most of us can't stand school."
> "They always ask me why I do all of the extra stuff for class but I tell them if they were half as interested as me they would be filling wasted time with science, not just science, but wonderful science."
> "Most students at my school don't care about school. They care about socializing and partying more than the grades they make. Most of the time I have to assess what I want in life to motivate myself to do well."

Some young women responded to the question about peer support in science with general comments about how their *friends do not support them in science*:

> "They were too busy to encourage, not that they didn't want to."
> "They did nothing."

"No one was concerned about your progress in science."

"No one tried to encourage me or discourage me."

"Science just isn't a big deal to everyone like me, we all have our differences."

"I haven't really had a student come to me and talk about science. I know there are some kids in my school who truly like science but no one has encouraged me in science."

Some of the young women suggest that they *don't talk about science with their friends*:

"We have other things to talk about than science."

"It is never brought up in conversations."

"Actually we do not discuss science because everyone does not like it."

Some suggest that *they don't talk about school work in general*:

"We don't talk about school."

"We usually don't talk about school a lot. Most of us can't stand school."

Others suggest that *science is not high priority* in their peer groups:

"Normally my peers see me in different areas of work. Not a scientist but a lawyer, writer, entrepreneur."

"Never was a priority."

"The students have more things on their mind than just science."

"I had few friends and they were at least as messed up as I was. Encouraging each other in science was not tops on our list of priorities."

Related to this low priority and lack of discussion about science is the general notion that *the young women's friends are just not interested in science*:

"None of my friends are interested."

"I myself do not know anybody who wants to be in science."

Many of these comments on friends not being interested in science have a gendered aspect to them. It appears that they are suggesting their *friends do not like science because they are girls*:

"Girls like to look cute so they don't like to get into that stuff."

"Most females I hang out with, they always talk about how nasty science is."

"I think girls my age tend to think more about boys and cosmetics and making themselves look pretty."

Another theme in this section is that the young women's *friends do not do well in science and are discouraged or do not see a future in science.*

"Well, sometimes I felt stupid in comparison to the other students especially in math and science."

"Nobody really paid much attention to a career that involved the scientific methods."

"A lot of my friends want to make quick money and they don't see it happening in science."

In fact, some *young women and their friends were interested in science but became discouraged and stopped taking it:*

"I have several friends that went to school for computer science, they have degrees but were not able to find jobs in that field."

"I saw a lot of interest in some of the girls as far as science in high school. We had this program called SECME that promoted Science, Engineering, and Math. There were lots of girls in that program. I think that by the time they got to college, they might have changed their minds."

"Most of my friends were interested in grade school. But when we reached high school they said math and science was too hard so they went into other things like business and teaching."

"They too were discouraged."

Additionally, some note that there is a *general sense of student apathy, and that students are competitive and only interested in themselves.*

Some comments on *the apathy about science (and school)* and the lack of support in science among students include:

"Other kids thought I was smart and did not feel I needed any encouragement from them and some simply did not care."

"They didn't care."

"They did nothing … just glad for class to be over."

"Nobody cared if I liked science or not."

"At my school it was more like a 'to each his own' attitude."

"No one really cared about it."

"To some of them it was just another class."

"It did not concern them."

Some comments on the lack of peer support in science focusing on *fellow students who are only interested in themselves* include:

"They were concerned about their own situations."
"Everyone [was out for] themselves."
"They were thinking about themselves."
"They did not involve themselves in other people's affairs."
"Everybody was just worried about passing the class."
"[There was a] Lack of concern as most students were just trying to make it through the day."
"They do not involve themselves in other people's affairs."
"The only thing that they were concerned with was their own interest."

Similarly, one young woman commented on the lack of peer support due to the *competitive nature of the school context*:

"They want you to fail so that they can be on top."

In fact, some young women go so far as to say *they don't want support or advice from their peers*:

"I didn't really want my friends to encourage me. I didn't like science."
"No one can tell me what I want to do or should do [with] my future … it's my life."

Related to this notion of not wanting peer support, several of the young women suggest that *it is not peers' job to support each other in science*:

"It is not the student's job and I was quiet in high school so I really didn't talk much."
"We don't encourage anyone in our school to do anything. That is their business and also the teacher's job. We just mind our business and do our own things."
"I wasn't relying on other students to encourage me in science. That's not their job. It was hard enough to encourage yourself."
"It wasn't their job, but they spoke of their own enthusiasm."

Friends are interested in science, talk about it, and provide support and encouragement in science.

The second general theme in the young women's responses is in contrast to the first. In these responses, the young women talk about friends who are interested in science, talk about it, and provide support and encouragement in science. Although there are a good number of responses of this nature, they are less frequent in number than are the comments in the previous category where friends do not provide support and encouragement in science. Quantitative data from the Knowledge Networks survey support this conclusion. In response to a closed-ended question asking about whether students at the respondent's high school tend to encourage them in science, 80% of the young African American women answered "no." This figure was also large (but significantly smaller) in the sample of young white women. Seventy percent of that group also reported that their friends at school did not encourage them in science. The qualitative data is important in showing that a good number of the young women do have positive experiences with their friends in the science realm. These insights go beyond a simple answer of yes or no to a question about friends' encouragement in science.

Some of the comments from young African American women whose *friends are interested in science* include:

"When we are in science class most of my classmates seem pretty interested."

"Science is not about which race likes it better, it is about doing what you like, e.g., a lot of African Americans that I know are into science."

"I was preparing myself to attend college so I took preparatory courses, my group of friends was very close—we formed study groups. Our discussions were very interesting. The things we did in class like dissecting, chemistry lab were just very interesting to me."

"I know plenty of girls at my school love science."

"I like science. I know lots of females at my school love it. We even had a science club when I was in fifth–seventh grade."

"I know a lot of people that are interested in science, most of my friends are."

"Lots of people in high school were very interested in science."

Not only did some young women report friends that were interested in science but they reported getting *support from their friends in science*.

"We helped each other, and just knew that everyone would reach their potential."

"They would explain to me or show me another way to understand it."

"[They encourage me] to further my education in physics."

"If I had trouble figuring out a formula someone would come over and help me."

"When I won 2nd place in the science fair everyone encouraged me
in the finals that were held at a local university and I received
lots of support and offers of help from my fellow classmates."
"We had to encourage each other to make good grades."
"We talked about trips we went on before and after the trip."
"By telling me I am smart and asking me for help when they need
it."
"If I am having a hard time with a chapter we will get together and
study to help each other with what we do not understand."
"By engaging in group discussions."
"We all studied together at times."
"We did labs in groups. We always helped each other out. No one
tried to run the group by themselves.
"Whenever we had to do projects everyone was excited about other
people's project offering hints and asking how you were doing."
"By showing support and acceptance."
"Because there was the incentive of going away on a science based
trip, my friends and I always supported each other. If one of us
went, we all had to go. In turn, we all had to make the grades to
qualify."

More specifically, some young women got *support from friends in the
science classroom*:

"I have some of my close friends in science and when we are in
groups we want our lab to turn out the best."
"We all studied together at times."

Some received support from "mostly family and close friends that were
in my science classes with me."

In a related way, some noted that they *talk with their friends about
science*:

"They are always talking to me."
"[They] tell about the great experiments that are done in class."

Finally, one young woman got support from her friends who saw science as a way to make a better future. Her friends told her to:

"Make a better future: go into science."

Race and peer groups work to influence science.

A third general theme in the young women's responses involves issues
involving race, peer groups, and science. Here, it is the issue of the white

peer groups that are perceived as dominating and excelling in the science classroom. Teachers are not the only ones that project messages involving race and science. Peer groups are also a powerful factor.

Some of the comments focus on *white student's advantages and negative perceptions about blacks in science which can result in a feeling of not belonging in science*:

> "Well if you can't relate to anyone, and you don't want everyone to cater to you, science class usually goes into groups and your white counterparts feel like you're looking for a ride ... do all the work, in short your ideas aren't considered."

> "First of all, they really don't talk about black scientists unless it's black history month, and if you go to a school where there's mostly whites in your class how are you supposed to feel like you don't belong?"

> "The main students that understood science well were white."

> "I am the only African American female in nearly all of my classes. It is only natural to feel uncomfortable occasionally, but that doesn't stop me from striving to achieve my goals."

> "There were very few, if any black students taking science classes in that field when I was in school. Also, science is often considered a difficult subject and it sometimes feels that mainstream culture has no expectations for us to achieve in that area."

Gender and peer groups work to influence science.

The relationship between gender, peer groups, and science is a fourth theme that comes through when the young women talk about peer influences on their science experiences. Some of the young women see the *science classroom as being male dominated* with the young men excelling and the young women feeling left out:

> "Most science teachers, especially male, are more open and helpful to males."

> "Boys are selected or encouraged to participate in experiments first."

> "I've had male teachers who often overlooked me in favor of male students. I was in a class with a lot of males, and I didn't feel welcomed."

> "A lot of males may take that [science] course, then the females feel different and uncomfortable."

> "Males see females as having a limit when it comes to science."

> "The [boys] mess with the girls and try to mess them up in what they're doing."

Others suggest that even *boys don't like science and in fact, girls might like it better than boys.*

> "Well at my school boys don't really like any subjects. Most of our boys have dropped out of school. To me girls are more interested in things like that."
> "Most guys in my science classes throughout high school were not as interested as girls were."
> "Mostly females finish school and pay attention to their work."

Peer cultures view science as dumb and something that is for geeks and nerds.

Finally, some of the young women (and their peers) think the peer culture that is associated with science is dumb, and something for geeks and nerds. Science does not have a positive reputation among their friends. Those that are interested in science are made fun of. Some of the comments reflecting *the image of science as dumb, boring, not fun* are:

> "They say science is dumb."
> "It's not fun."
> "They think science is boring."
> "[They] said "that's for book worms.""

Some of the comments that reflect the *negative stereotypes (held by peers) that label scientists as nerds and geeks* include:

> "Most think only geeks like the subject."
> "[They] tease people interested in science. Calling them 'nerds' and 'eggheads.'"
> "It's for nerds, many people think."
> "Anyone interested in science was seen as a geek or a nerd."

Conclusions

Peer groups are one of the strongest influences on young people during adolescence. The literature on African American peer groups presents a mixed message of peers who are discouraged and giving up on the system on the one hand, and peers with high educational aspirations and trust in the educational system on the other hand. The data from the NELS and Knowledge Networks peer analyses support the notion of complexity in the nature of peer influences for young African American women in the area of science.

Quantitative data from the NELS survey show that young African American women, overall, do not have disadvantages in the area of peer resources relative to young white women. In fact, they have advantages on a number of key peer resources including having friends who think it is important to study and friends who think it is important to get good grades. These peer resources often have important positive influences on experiences in science. However, the young African American women are slightly less likely than their white counterparts to have friends who think it is important to finish high school. Nevertheless, a large majority (80%) of the young African American women do have friends who value finishing high school and this variable is one of the most important peer influences on science experiences.

Qualitative Knowledge Networks data show a complex set of peer influences. Peer influences include the gendered aspect of science. The young women who have friends who do not feel interested or qualified in science often discuss this in gendered terms about girls and science. The young women's reflections on peer influences also include aspects of race. The negative label put on science and scientists (geeks and nerds) is also an important deterrent in some of the young women's peer groups. Throughout the comments, the difficulty of the school environment and the discouragement in school in general and science, in particular, is revealed. Thus, these structures involving educational systems, gender, and race come through as barriers in the young women's reflections on peers and science. However, the comments also suggest that peers do sometimes provide considerable agency. Although quantitative data from Knowledge Networks suggest that a larger number of young women report not receiving support from peers in science (relative to those receiving support), the qualitative data reveal considerable peer excitement and engagement and support in science among these young women. A good number of the young women report knowing a considerable number of women who like science at their school. In fact, a rather strong theme in the comments was the fact that it is young *men* who are not interested in school or science and who often drop out of school. Earlier in this chapter, I noted the considerable literature that reveals the difficulties of young men (especially young African American men) in contemporary U.S. education systems.

In contrast to the research by Ogbu and others on issues of race and achievement in the general educational system, we find no evidence that the young African American women are pressured by peers to do poorly in science in an effort to avoid acting white. However, we do find evidence that they see race as a barrier to doing well in science. Research focusing on African American students in general, or on African American men, is likely to produce very different results than that focusing on young African American women. The variety of data sets and methods used here are helpful in providing an in-depth picture revealing the complexities of young African American women's peer influences in science. The negative aspects of race and gender, along with negative images of scientists and discouragement (in general) in the school system, take away from science engagement for some. However, a good number of the young women are excited about science, have peers who are also excited about science, and perceive their peers as being supportive in this realm. They spoke of study groups, helping each other out, a shared love of science, excitement about science projects and experiments, and being encouraged in science in order to "make a better future."

7 Conclusions

*"Science Is Not About Which Race Likes It Better,
It Is About Doing What You Like."*

Women and minorities have increased their representation in science education and occupations (Hanson et al., 2004; National Science Foundation, 2004). However, the culture of science continues to be white and male. It is often hostile to women and minorities (Catalyst, Inc., 1992; Harding, 1986; Rossiter, 1982; National Science Foundation, 2000; Pearson et al., 1999; Ramirez and Wotipka, 2001).

Research on women in science has proliferated, but the focus is often on differences between men and women with little attention to subgroups of women. It is a mistake to think of women as an undifferentiated group. In fact, preliminary research has suggested that young African American women are particularly interested in science (sometimes more so than their white counterparts) (Hanson and Palmer-Johnson, 2000; Hanson, 2004; National Center for Education Statistics, 2000a). In spite of this interest, African American women experience racism and sexism in the science domain and remain underrepresented in science programs and occupations (Malcom et al., 1998; Vining Brown, 1994; National Science Board, 2000). The paradox at the center of this book involves the interest and engagement in science among many young African American women in the context of a science culture that remains openly hostile to those who are not white and not male.

The conceptual framework that I use to unravel the complex interaction between race and gender in the science domain is one that focuses on gender and race as structures that organize our lives and often contribute to inequality. It is not so much qualifications but these master statuses of gender and race that determine who enters and succeeds in science. The framework is multicultural and feminist in suggesting that gender structures are powerful aspects of organization but they are not identical across cultures. It is important that diversity in gender systems across race/ethnic groups is acknowledged. The framework utilizes a critical perspective and argues that it is not structure alone, that creates and discourages opportunities in science. Women and African Americans and African American women are not merely "victims" of racism and sexism. Sometimes structures and status quos are questioned and challenged. Thus, it is not just structure that is important for understanding African American women's experiences in science, but also these young women's responses to race and gender structures.

The goal of the book has been to examine the experiences of young African American women in science education. What are their attitudes about science, what science courses do they take, and do they succeed in the science realm? Additionally, I examined the extent to which these young women saw discrimination and a chilly climate in the science classroom. A second goal of the book involved an examination of young African American women's experiences in school systems, families, communities, and peer groups. I have examined how each of these contexts works to encourage and discourage the young women's interest and achievement in science. The major research focus has been the experiences of young African American women in science and the push and pull factors in a variety of areas of life that influence these experiences.

The research questions were addressed using a triangulation of methods and data sets. An analysis of the National Educational Longitudinal Study (NELS) of American youth lent insight into the experiences of a large, nationally representative sample of high school students in the science curriculum. The data included rich information on science experiences, as well as experiences involving schools, families, communities and peers and thus allowed measurement and understanding of the processes at work in young African American women's science experiences.

Information from the secondary NELS data was supplemented by data from the Web survey of young African American women. The survey was designed by the author and conducted in cooperation with Knowledge Networks in 2003. The Web survey attempted to gain further insight into young African women's science experiences by using vignettes and qualitative data from open-ended questions. It utilized a new Web technology that combined probability sampling with the reach and capabilities of the Internet. The open-ended questions (as well as focused closed-ended questions) inquired specifically about the variety of influences in schools, family, community, and peer groups that encourage or discourage young women in science education. As Feagin et al. (1996) argue, it is critical that we allow minorities to explain their experiences in their own words.

Our Web survey was unique in utilizing the vignette technique to better understand young African American women's perceptions of discrimination in the science classroom. Although the attitude survey is the standard technique in social science research, questionnaires and surveys are not always well suited for the study of human attitudes and behaviors. They can produce unreliable and biased self-reports (Alexander and Becker, 1978). Standardized surveys often use vague questions that are interpreted within the respondent's own meaning system. One solution, then, is to make the stimulus given to the respondent as concrete and detailed as possible, as in a vignette. This stimulus should resemble a real-life decision making or judgment-making situation. In addition, since the stimulus can be held constant or varied in a systematic fashion, it gives the researcher greater control, as in experimental designs. The vignette is an ideal strategy for gaining this detail and control.

In Chapter 3, I described young African American women's experiences in science using data from NELS and the Knowledge Networks Web survey. Some have argued a double disadvantage for African American women in science (Clewell and Anderson, 1991; Cobb, 1993; Vining Brown, 1994). My findings do not support this. Many young African American women were found to be fascinated with science and planning on careers in science. Yet they do not feel welcome. *Over one-half* of young African American women in a nationally representative sample say that African American women *do not feel welcome in science.*

When considered as a yes/no question in the Knowledge Networks survey, only one-third of the young women reported that they like science. This question was followed with an open-ended item, "Why do you like, or not like, science?" The answers to this question suggested that many of the young women actually have mixed feelings about science. One of the lessons learned from this finding is that young people's feelings toward science are not reflected in a single question on how much they like science. Likes and dislikes are complex things that seldom follow simple dichotomies. Researchers interested in measuring attitudes toward science need to allow young people to reflect this complexity in their responses. If researchers use quantitative measures it is important that they include questions about what they like about science and what they dislike about science without simply asking whether they like or dislike science.

Chapter 4 examined the role of schools in young African American women's science experiences. Literature suggests that most parents of minority students have a high value on education, high aspirations for their children, and a desire for involvement in their children's academic development (Cummins, 1993). Student's values mirror these (Ainsworth-Darnell and Downey, 1998). However, school systems in the United States have race and social class biases that limit the achievement of many young people in minority statuses. The past decades have seen improvement, but many biases (some subtle and some not so subtle) still exist (Banks, 1991; Persell, 1977; Kao and Thompson, 2003; Davis, 2004; Feagin et al., 1996). African American youth are aware of this structured disadvantage in the educational system and in the broader society (Ogbu, 1978, 1985, 1991; Davis, 2004; Feagin et al., 1991). For many (especially in higher education), racism is a central component of their education experience (Feagin et al., 1991). Similar barriers exist in science education systems. Early work on science as fair has not been supported (Cole and Cole, 1973; Cole, 1987). Although underrepresented, African Americans have made major contributions to science (Ovelton Sammons, 1990). However, many are unaware of these contributions since teachers, curriculum, and textbooks often overlook minority scientists and their discoveries (Pearson, 1985).

Findings on school effects in the NELS data show that the young white women have an advantage on variables measuring teacher interest, school honors, and school programs. They are more likely than their African American counterparts to report teachers that are interested in them, to have won an academic honor, to have received recognition for good grades, to be in a private school, and to be in a college prep program at their school. Findings show that these school factors are critical for success in science.

The young women's responses to open-ended questions in the Knowledge Networks Web survey provided a wealth of information about their experiences in the school system and in the science classroom. The young women are often discouraged by lack of preparation, isolation, lack of diversity, discrimination, and poor teaching. Yet they also have considerable agency when it comes to their experiences with science. Importantly, they are observing and thinking about their experiences and are able to reflect on the nature of the science classroom, how it affects them, and how it can be improved. Their thoughts often provide insights into the way in which schools and teachers veer them away from science and lower their confidence. They also provide insights into how schools can change as well as the need for an equal footing in science.

In Chapter 5, I focused on the context of the African American family and community. My research has shown the considerable interest and involvement of young African American women in the science domain, in spite of the fact that teachers and schools often do not see them as science talent. Although negative stereotypes about African American families are abundant, my findings suggest these families are often a source of strength and resilience in the educational realm (especially for young women). African American communities, in general, are as a source of community capital given their focus on community and their investment in children, especially females (in contrast to the individual focus and greater investment in males typical of white communities). Community investment in children results in young people's desire (and obligation) to give back to the community through achievement in elite areas such as science. Success is seen as a group effort that involves support of family and community.

Data from the NELS survey show that young white women may have more socioeconomic resources than African American women, but the reverse is often the case for family variables measuring parent's educational expectations and family involvement. Additionally, although there are no race differences on some community variables, others such as peer value systems involving community work and religious activities as well as respondent's involvement in religious activities are reported with more frequency among the young African American women (relative to the white women). Many of these family and community characteristics are shown to be related to success in science in the analyses.

Qualitative data from the Knowledge Networks survey show a considerable feeling of independence among the young African American women. They feel that they make their own decisions without the influence of others. This independence may be critical for survival in educational systems that are less than welcoming. The qualitative data also allow the young women to report their thoughts and experiences about their families in their own words. Here we can read in report after report about mothers who were the young woman's biggest cheerleader, families full of women who were interested in science, fathers who hung up posters of African American scientists, and mothers who let them use kitchen utensils to dig up worms. In spite of this support, many of the young women felt discouraged in science. As one young woman reported, "He [my father] actually tried very hard [to support my science interests], but it was kind of him swimming against the tide of the world."

Chapter 6 examines the role of peers in young African American women's science experiences. Data from the NELS survey show that young African American women, overall, do not have disadvantages in the area of peer resources relative to young white women. In fact, they have advantages on a number of key peer resources including having friends who think it is important to study and friends who think it is important to get good grades. These peer resources often have important positive influences on experiences in science. However, the young African American women are less likely than are their white counterparts to have friends who think it is important to finish high school and this variable is one of the most important peer influences on science experiences.

Qualitative Knowledge Networks data show a complex set of peer influences. Peer influences include the gendered aspect of science. The young women who have friends who do not express interest or qualification in science often discuss this in gendered terms about girls and science. The young women's reflections on peer influences also include aspects of race. The negative label put on science and scientists (geeks and nerds) is also an important deterrent in some of the young women's peer groups. Throughout the comments, the difficulty of the school environment and the discouragement with school in general and science in particular, is revealed. Thus, these structures involving educational systems, gender, and race come through as barriers in the young women's reflections on peers and science. However, the comments also suggest that peers do sometimes provide considerable agency. Although a closed-ended question from the Knowledge Networks survey suggests that a majority of young African American women report *not* receiving support from peers in science, the qualitative data reveal considerable peer excitement, engagement and support in science among these young women. A good number of the young women report knowing a considerable number of women who like science at their school.

In contrast to the research by Ogbu (1974) and others (e.g., Fordham and Ogbu, 1986) on issues of race and achievement in the general educational system, no evidence that the young African American women are pressured by peers to do poorly in science in an effort to avoid acting white is found. However, there is evidence that race is seen as a barrier to doing well in science. Research focusing on African American students in general, or on African American men, is likely to produce very different results than that focusing on young African American women. The variety of data sets and methods used here are helpful in providing an in-depth picture revealing the complexities of young African American women's peer influences in science. The negative aspects of race and gender, along with negative images of scientists and discouragement (in general) in the school system, take away from science engagement for some. However, a good number of the young women are excited about science, have peers who are also excited about science, and perceive their peers as being supportive in this realm. They spoke of study groups, helping each other out, a shared love of science, excitement about science projects and experiments, and being encouraged in science in order to "make a better future."

The paradox of young African American women's interest in science in the context of a chilly climate for those who are not white and not male is a complex one that requires a complex explanation. I hope that the research in this volume has contributed to unraveling this complexity. The literature and analyses presented here within the context of the critical feminist framework present a picture of a school system that does not tend to see young African American women as science talent. There is evidence of some teachers who make a difference. However, overall, the science education system is structured in a way that favors white, middle-class males. The answer to the paradox continues to evolve as the African American family, community, and peer systems are examined. As the work on agency suggests, sometimes events and circumstances come together in unexpected ways. The African American family and community is seen here as one that has a different definition of femininity than in the white family and community. African American women have historically been viewed as strong individuals who work and head families. Gender is often thought of in a universal way—especially within a particular society. Evidence is provided here of a unique gender system in the African American family, community, and peer system that sometimes works against all odds to encourage interest and activity in science. The processes revealed by the NELS and Web survey data allow the reader to see the complexity and subtleties of the push and pull factors that influence young African American women in the science domain. It is when we allow these young women to provide answers in their own words, that the paradox of young African American women's success in science (in spite of the "tide" that they must swim against) is unraveled.

Appendix A: Tables

TABLE A.3.1   Means (and Standard Deviations) on Science Variables for Young Women by Race: NELS

	Race	
	African American Females	White Females
Science Access		
1988 (8th Grade)		
In advanced, enriched or accelerated science courses (1 = yes; 0 = no)	.40 (.49)*	.23 (.42)*
1990 (10th Grade)		
Coursework in chemistry (1 = one half to two years; 0 = none)	.18 (.38)	.18 (.38)
1992 (12th Grade)		
Enrolled in science classes in the last 2 years (1 = yes; 0 = no)	.89 (.32)	.89 (.31)
1994		
Science major at postsecondary institution† (1 = yes; 0 = no)	.25 (.43)*	.14 (.34)*
2000		
First postsecondary degree—science‡ (1 = yes; 0 = no)	.14 (.35)	.14 (.34)
Science degree expected by age 30 (1 = yes; 0 = no)	.26 (.43)*	.15 (.36)*

TABLE A.3.1 *Continued*

	Race	
	African American Females	White Females
Science Achievement		
1988 (8th Grade)		
Science grades from 6th grade until now (1 = mostly A's ; 0 = B's or less)	.34 (.47)*	.38 (.49)*
Science Standardized Score	45.12 (7.91)*	53.73 (9.11)*
1990 (10th Grade)		
Science grades (1 = mostly A's; 0 = B's or less)	.38 (.48)*	.43 (.50)*
Science Standardized Score	43.97 (7.09)*	52.06 (8.98)*
1992 (12th Grade)		
Science Standardized Score	43.02 (8.12)*	52.10 (8.87)*
2000		
Current/most recent occupation—science (1 = yes; 0 = no)	.22 (.42)*	.17 (.37)*
Science Attitudes		
1988 (8th Grade)		
Usually look forward to science class (1 = strongly agree or agree; 0 = disagree or strongly disagree)	.59 (.49)	.55 (.50)
Afraid to ask questions in science class (1 = yes; 0 = no)	.86 (.34)	.87 (.34)
Science will be useful in my future (1 = strongly agree or agree; 0 = disagree or strongly disagree)	.70 (.46)*	.65 (.48)*
1990 (10th Grade)		
Often work hard in science class§ (1 = almost everyday; 0 = less than everyday)	.64 (.48)*	.55 (.50)*
1992 (12th Grade)		
Interested in science§ (1 = important; 0 = not as important)	.42 (.50)	.49 (.50)
Does well in science (1 = important; 0 = not as important)	.48 (.50)	.50 (.50)
Need science for job after high school (1 = yes; 0 = no)	.30 (.46)*	.17 (.38)*

TABLE A.3.1 *Continued*

	Race	
	African American Females	White Females
Science Attitudes (*continued*)		
2000		
Planned occupation by age 30—science (1 = yes; 0 = no)	.31 (.46)*	.24 (.43)*

* T-test for difference in means significant at .05 level.

† Questions only asked of those in college at time of interview.

‡ Questions only asked of those who had attended or were attending college at time of interview.

§ Questions only asked of those taking science.

TABLE A.3.2 Means (and Standard Deviations) on Science Outcomes for African American and White Women: Knowledge Networks Data†

	Race	
Science Outcomes	African American	White
Expect science occupation age 30 (1 = yes; 0 = no)	.28 (.45)*	.21 (.41)*
Hope for science occupation at age 30 (1 = yes; 0 = no)	.29 (.46)*	.22 (.42)*
High school grades science (6 = A's; 1 = below C)	3.80 (1.35)*	4.36 (1.30)*
Good in science (5 = very good; 1 = not so good)	2.93 (1.09)*	3.15 (1.09)*
Like science (5 = very good; 1 = not so good)	2.85 (1.30)*	3.19 (1.22)*
Feel welcome in science (1 = yes; 0 = no)	.66 (.48)*	.80 (.40)*
R's science interests influenced by others (1 = yes; 0 = no)	.39 (.49)*	.51 (.50)*
Race a bigger barrier in science than gender (1 = yes; 0 = no)	.48 (.50)*	.35 (.49)*

* T-test for differences in means significant at .05 level.

† R = respondent.

TABLE A.3.3 Multiple Classification Results Showing Means (and Deviations from Sample Means) on Science Variables for Young Women by Type of Vignette: Knowledge Networks Data†

	Science Outcomes	
	Has This Ever Happened to You	Others Like Woman in Vignette Don't Feel Welcome in Science
Vignettes		
A. Sample: African American Women		
1. Girl in Vignette: African American		
• Race as issue	.31 (.06)	.52 (.12)
• Gender as issue	.16 (−.08)	.53 (.14)
• Neutral	.53 (.29)	.63 (.23)
2. Girl in Vignette: White		
• Gender as issue	.21 (−.03)	.35 (−.05)
B. Sample: White Women		
1. Girl in Vignette: African American		
• Race as issue	.16 (−.07)	.36 (−.04)
• Gender as issue	.18 (−.06)	.36 (−.03)
• Neutral	.36 (.11)	.39 (−.01)
2. Girl in Vignette: White		
• Gender as issue	.18 (−.06)	.31 (−.08)
Mean across groups	.24	.40
F	9.95*	5.03*

* Anova model is significant at .05 level.

† Higher score indicates greater support for the student in vignette, which in general measures problems/discomfort in science.

TABLE A.4.1 Means (and Standard Deviations) on School Variables for
African American and White Women: NELS

	Race	
School Variables	African American	White
Teaching is good (1 = agree; 0 = disagree)	.83 (.37)*	.87 (.34)*
Teachers interested in students (1 = agree; 0 = disagree)	.77 (.42)*	.82 (.38)*
Importance of teachers in science decision (1 = very; 0 = somewhat, not)	.32 (.47)	.28 (.45)
Importance of counselor in science decision (1 = very; 0 = somewhat, not)	.47 (.50)*	.32 (.47)*
Real school spirit at school (1 = agree; 0 = disagree)	.68 (.47)	.70 (.46)
Students friendly to other racial groups (1 = agree; 0 = disagree)	.90 (.30)*	.81 (.39)*
Doesn't feel safe at school (1 = agree; 0 = disagree)	.16 (.37)*	.07 (.26)*
Won an academic honor (1 = yes; 0 = no)	.18 (.39)*	.25 (.43)*
Received recognition for good grades (1 = yes; 0 = no)	.49 (.50)*	.55 (.50)*
School type (1 = private; 0 = public)	.06 (.23)*	.10 (.29)*
High school program (1 = college prep; 0 = other)	.39 (.49)*	.49 (.50)*

* T-test for difference in means significant at .05 level.

TABLE A.4.2 Means (and Standard Deviations) on Science Outcomes for African American Women by School Characteristics: NELS

School Variables	Access					
	1988 (8th Grade) In Advanced, Enriched, or Accelerated Science Course	1990 (10th Grade) Coursework in Chemistry	1992 (12th Grade) Enrolled in Science Classes in the Last Two Years	1994 Science Major at First Postsecondary Institution	2000 First Postsecondary Degree—Science	2000 Science Degree Expected by Age 30
Teaching is good						
Agree	.39 (.49)**	.21 (.41)**	.88 (.33)*	.25 (.43)**	.14 (.35)	.18 (.39)**
Disagree	.24 (.43)**	.06 (.24)**	.94 (.24)*	.11 (.32)**	.12 (.34)	.39 (.49)**
Teachers interested in students						
Agree	.37 (.49)	.18 (.39)	.87 (.84)**	.28 (.45)**	.14 (.35)	.21 (.41)**
Disagree	.40 (.49)	.24 (.43)	.94 (.24)**	.08 (.28)**	.13 (.34)	.04 (.20)**
Importance of teachers in science decision						
Very	.60 (.50)**	.32 (.47)**	†	.39 (.50)	.35 (.49)**	.26 (.45)
Not, Some	.36 (.48)**	.14 (.35)**		.35 (.48)	.08 (.28)**	.34 (.48)
Importance of counselor in science decision						
Very	.47 (.50)*	.19 (.40)	†	.29 (.46)	.24 (.44)	.32 (.47)
Not, Some	.33 (.47)*	.19 (.40)		.40 (.50)	.18 (.39)	.34 (.48)
Real school spirit at school						
Agree	.39 (.49)*	.19 (.39)	.89 (.31)	.22 (.42)	.13 (.33)	.17 (.38)**
Disagree	.31 (.46)*	.19 (.39)	.89 (.32)	.21 (.41)	.18 (.38)	.33 (.47)**
Students friendly to other racial groups						
Agree	.35 (.48)*	.19 (.39)	.89 (.32)	.22 (.42)	.13 (.34)	.23 (.42)
Disagree	.48 (.51)*	.18 (.39)	.93 (.27)	.22 (.42)	.23 (.43)	.15 (.37)

Doesn't feel safe at school						
Agree	.48 (.50)**	.29 (.46)**	.96 (.19)**	.09 (.29)**	.04 (.21)*	.25 (.44)
Disagree	.35 (.48)**	.17 (.38)**	.87 (.33)**	.25 (.43)**	.15 (.36)*	.25 (.41)
Won an academic honor						
Yes	.48 (.50)**	.37 (.49)**	.95 (.22)**	.34 (.48)**	.25 (.44)**	.19 (.40)
No	.36 (.48)**	.16 (.37)**	.87 (.33)**	.21 (.41)**	.10 (.31)**	.25 (.43)
Received recognition for good grades						
Yes	.44 (.50)**	.30 (.46)**	.95 (.23)**	.24 (.43)	.17 (.37)*	.26 (.44)
No	.32 (.47)**	.09 (.29)**	.83 (.38)**	.22 (.42)	.10 (.30)*	.20 (.41)
School type						
Private	.41 (.50)	.14 (.36)	.89 (.32)	.24 (.44)	.09 (.29)	.21 (.41)
Public	.40 (.49)	.19 (.39)	.89 (.32)	.24 (.44)	.15 (.35)	.26 (.44)
High school program						
College Prep	.48 (.50)**	.38 (.49)**	.99 (.10)**	.28 (.45)	.18 (.39)**	.22 (.42)
Other	.34 (.48)**	.05 (.23)**	.82 (.39)**	.22 (.42)	.09 (.29)**	.28 (.45)

TABLE A.4.2 *Continued*

School Variables	Achievement					
	1988 (8th Grade) Science Grades from 6th Grade Until Now (Mostly A's or A's & B's)	1988 (8th Grade) Science Standardized Score (Top Quartile)	1990 (10th Grade) Science Grades (Mostly A's or A's & B's)	1990 (10th Grade) Science Standardized Score (Top Quartile)	1992 (12th Grade) Science Standardized Score (Top Quartile)	2000 Current/Most Recent Occupation—Science
Teaching is good						
Agree	.33 (.47)**	.07 (.25)	.37 (.49)**	.07 (.24)	.07 (.25)	.22 (.45)
Disagree	.21 (.41)**	.04 (.19)	.51 (.50)**	.06 (.23)	.04 (.19)	.28 (.41)
Teachers interested in students						
Agree	.30 (.46)**	.07 (.26)	.36 (.48)	.07 (.26)	.07 (.26)	.19 (.40)
Disagree	.44 (.50)**	.04 (.19)	.43 (.50)	.04 (.21)	.04 (.19)	.23 (.42)
Importance of teachers in science decision						
Very	.29 (.46)	.10 (.31)	.45 (.50)**	.07 (.25)	.10 (.31)	.28 (.46)
Not, Some	.35 (.48)	.16 (.37)	.53 (.50)**	.23 (.42)	.16 (.37)	.31 (.47)
Importance of counselor in science decision						
Very	.34 (.48)	.15 (.36)	.36 (.49)**	.08 (.28)	.15 (.36)	.16 (.37)**
Not, Some	.34 (.48)	.12 (.33)	.55 (.50)**	.15 (.36)	.12 (.33)	.35 (.48)**
Real school spirit at school						
Agree	.31 (.46)	.09 (.28)	.42 (.49)**	.05 (.21)*	.05 (.22)*	.22 (.42)
Disagree	.33 (.47)	.13 (.34)	.34 (.48)**	.10 (.30)*	.08 (.27)*	.23 (.42)
Students friendly to other racial groups						
Agree	.31 (.48)	.06 (.23)	.40 (.49)	.06 (.24)*	.06 (.23)	.23 (.42)
Disagree	.33 (.46)	.11 (.31)	.31 (.47)	.11 (.32)*	.11 (.31)	.19 (.40)

Doesn't feel safe at school						
Agree	.35 (.48)	.00 (.00)**	.20 (.40)**	.03 (.18)	.00 (.00)**	.36 (.18)**
Disagree	.31 (.46)	.07 (.26)**	.43 (.50)**	.07 (.26)	.07 (.26)**	.20 (.40)**
Won an academic honor						
Yes	.58 (.43)**	.23 (.42)**	.54 (.50)**	.20 (.41)**	.23 (.42)**	.29 (.46)
No	.25 (.50)**	.03 (.17)**	.33 (.50)**	.04 (.19)**	.03 (.17)**	.23 (.42)
Received recognition for good grades						
Yes	.46 (.50)**	.10 (.30)**	.51 (.50)**	.11 (.32)**	.10 (.30)**	.14 (.34)
No	.16 (.37)**	.03 (.16)**	.21 (.41)**	.03 (.16)**	.03 (.16)**	.14 (.35)
School type						
Private	.35 (.49)	.15 (.36)	.42 (.50)	.26 (.45)**	.14 (.35)*	.33 (.48)*
Public	.34 (.48)	.10 (.30)	.37 (.48)	.05 (.22)**	.06 (.23)*	.22 (.41)*
High school program						
College Prep	.51 (.50)**	.11 (.31)**	.42 (.50)*	.14 (.34)**	.11 (.31)**	.29 (.46)**
Other	.21 (.41)**	.02 (.16)**	.35 (.48)*	.01 (.10)**	.02 (.16)**	.18 (.39)**

TABLE A.4.2 *Continued*

				Attitudes			
School Variables	1988 (8th Grade) Usually Look Forward to Science Class	1988 (8th Grade) Science Will Be Useful in My Future	1990 (10th Grade) Often Work Hard in Science Class	1992 (12th Grade) Interested in Science	1992 (12th Grade) Does Well in Science	1992 (12th Grade) Need Science for Job After High School	2000 Planned Occupation by Age 30—Science
Teaching is good							
Agree	.60 (.50)**	.74 (.40)**	.67 (.47)	.45 (.50)**	.51 (.47)*	.31 (.46)	.28 (.50)**
Disagree	.43 (.49)**	.44 (.50)**	.66 (.48)	.17 (.38)**	.29 (.50)*	.26 (.47)	.44 (.49)**
Teachers interested in students							
Agree	.63 (.48)**	.70 (.46)*	.67 (.47)*	.47 (.50)**	.53 (.50)**	.34 (.48)**	.32 (.47)**
Disagree	.47 (.50)**	.78 (.42)*	.60 (.49)*	.12 (.34)**	.21 (.42)**	.12 (.33)**	.13 (.34)**
Importance of teachers in science decision							
Very	.64 (.49)	.80 (.41)	.54 (.51)	.49 (.51)	.68 (.47)**	.47 (.51)**	.52 (.51)**
Not, Some	.62 (.49)	.73 (.45)	.61 (.49)	.40 (.49)	.37 (.49)**	.23 (.42)**	.63 (.49)**
Importance of counselor in science decision							
Very	.64 (.49)	.75 (.44)	.62 (.49)	.32 (.47)*	.50 (.51)	.34 (.45)*	.23 (.43)*
Not, Some	.56 (.50)	.72 (.45)	.58 (.50)	.48 (.32)*	.45 (.50)	.17 (.38)*	.41 (.50)*
Real school spirit at school							
Agree	.59 (.49)	.70 (.45)**	.67 (.47)	.44 (.50)	.50 (.50)	.35 (.48)*	.29 (.46)
Disagree	.55 (.50)	.63 (.49)**	.65 (.48)	.37 (.49)	.42 (.50)	.18 (.39)*	.34 (.48)
Students friendly to other racial groups							
Agree	.57 (.50)	.68 (.47)	.63 (.49)	.42 (.50)	.49 (.50)	.29 (.46)	.32 (.47)*
Disagree	.60 (.50)	.72 (.45)	.67 (.47)	.40 (.50)	.40 (.51)	.41 (.52)	.20 (.40)*

Doesn't feel safe at school							
Agree	.65 (.48)	.63 (.49)	.52 (.50)**	.41 (.51)	.68 (.50)*	.28 (.47)	.17 (.38)**
Disagree	.56 (.50)	.70 (.40)	.69 (.46)**	.42 (.50)	.48 (.48)*	.31 (.46)	.33 (.47)**
Won an academic honor							
Yes	.65 (.48)	.75 (.44)*	.62 (.49)	.59 (.50)**	.62 (.49)*	.36 (.49)	.39 (.49)
No	.58 (.49)	.66 (.48)*	.66 (.47)	.36 (.48)**	.43 (.50)*	.29 (.46)	.31 (.47)
Received recognition for good grades							
Yes	.61 (.49)	.68 (.47)	.59 (.49)**	.52 (.50)**	.58 (.50)**	.39 (.49)*	.30 (.46)
No	.57 (.50)	.67 (.47)	.71 (.46)**	.32 (.47)**	.38 (.49)**	.22 (.42)*	.35 (.50)
School type							
Private	.79 (.70)	.73 (.45)	.65 (.48)	.55 (.51)	.42 (.51)	.62 (.50)**	.51 (.51)**
Public	.59 (.59)	.71 (.46)	.65 (.48)	.40 (.49)	.49 (.50)	.26 (.44)**	.29 (.46)**
High school program							
College Prep	.67 (.47)**	.77 (.42)**	.55 (.50)**	.49 (.50)**	.51 (.50)	.39 (.49)**	.31 (.47)
Other	.54 (.50)**	.66 (.48)**	.76 (.43)**	.31 (.47)**	.44 (.50)	.16 (.38)**	.30 (.46)

* T-Test for difference in means significant at .20 level.

** T-Test for difference in means significant at .05 level.

† In some situations the difference in means test was not possible due to small N.

TABLE A.4.3 Results from Logistic Regression Models Showing Effects of School Variables on Science Outcomes for African American Women: NELS†

School Variables	Access				Achievement		Attitudes
	1992 (12th Grade) Enrolled in Science Classes in the Last Two Years	1994 Science Major at First Postsecondary Institution	2000 First Postsecondary Degree—Science	2000 Science Degree Expected by Age 30	1992 (12th Grade) Science Standardized Score (Top Quartile)	2000 Current/Most Recent Occupation—Science	2000 Planned Occupation by Age 30—Science
Teaching is good (1 = agree; 0 = disagree)	.49 (.71)	1.14 (.50)	2.36 (.79)	1.20 (.59)	1.71 (1.08)	3.08 (.48)**	.51 (.44)*
Teachers interested in students (1 = agree; 0 = disagree)	.68 (.64)	2.02 (.50)*	.71 (.66)	4.36 (.85)*	1.36 (.97)	.60 (.37)*	2.16 (.44)*
Real school spirit at school (1 = agree; 0 = disagree)	1.26 (.39)	.94 (.15)	1.08 (.48)	.90 (.41)	.54 (.28)	1.36 (.31)	1.46 (.31)
Students friendly to other racial groups (1 = agree; 0 = disagree)	.34 (.77)*	.92 (.49)	.53 (.71)	1.52 (.79)	.53 (.77)	1.07 (.45)	1.85 (.52)
Doesn't feel safe at school (1 = agree; 0 = disagree)	5.55 (.78)**	.24 (.64)**	.07 (1.39)**	2.64 (.59)*	.00 (‡)	2.94 (.36)**	.54 (.41)*
Won an academic honor (1 = yes; 0 = no)	.97 (.66)	1.46 (.36)	2.08 (.48)*	1.07 (.48)	5.28 (.58)**	1.05 (.35)	2.95 (.38)**
Received recognition for good grades (1 = yes; 0 = no)	1.55 (.44)	1.02 (.33)	1.24 (.50)	.61 (.41)	1.33 (.66)	2.29 (.31)**	.38 (.32)**

School type (1 = private; 0 = public)	.41 (.83)	.77 (.53)	.34 (.88)	.88 (.64)	1.39 (.76)	1.82 (.47)	3.63 (.53)**
High school program (1 = college prep; 0 = other)	25.76 (.81)**	1.72 (.32)*	2.12 (.46)*	1.64 (.39)	1.47 (.61)	1.31 (.30)	1.44 (.29)*
SES	1.65 (.20)**	.94 (.15)	1.12 (.20)	1.07 (.18)	2.07 (.28)**	1.06 (.13)	.68 (.14)**
Constant	6.81 (1.02)*	.17 (.81)**	.07 (1.11)**	.03 (1.22)**	.01 (1.26)**	.04 (.70)**	.39 (.68)*
Model χ^2 (df)	74.40 (10)**	19.99 (10)**	19.40 (10)**	9.58 (10)	49.62 (10)**	41.23 (10)**	40.45 (10)**

* Significant at .20 level.
** Significant at .05 level.
† Table shows antilogs and standard errors in parentheses.
‡ Large standard error here.

TABLE A.4.4 Results from Logistic Regression Models Showing Effects of School, Family, Community, and Peer Variables on Science Outcomes for African American Women: NELS†

	Access				Achievement		Attitudes
	1992 (12th Grade) Enrolled in Science Classes in the Last Two Years	1994 Science Major at First Postsecondary Institution	2000 First Postsecondary Degree—Science	2000 Science Degree Expected by Age 30	1992 (12th Grade) Science Standardized Score (Top Quartile)	2000 Current/Most Recent Occupation—Science	2000 Planned Occupation by Age 30—Science
School Variables							
Teachers interested in students (1 = agree; 0 = disagree)	.53 (.67)	1.17 (.50)	.52 (.67)	7.00 (.99)**	8.17 (1.66)*	1.94 (.48)*	1.44 (.46)
Won an academic honor (1 = yes; 0 = no)	5.58 (1.15)*	2.24 (.38)**	2.76 (.54)*	.71 (.52)	3.32 (.64)*	1.98 (.38)*	2.36 (.40)**
High school program (1 = college prep; 0 = other)	9.97 (.83)**	1.14 (.75)	1.55 (.54)	1.67 (.46)	2.23 (.74)	1.29 (.34)	1.29 (.33)
Family Variables							
Mother's occupation (1 = professional; 0 = other)	.41 (.70)	1.25 (.53)	.19 (1.10)*	.47 (.66)	3.53 (.75)*	1.85 (.48)*	1.02 (.53)
Father's occupation (1 = professional; 0 = other)	14.48 (2.50)	.67 (.51)	.32 (.85)*	2.52 (.54)*	2.01 (.69)	.95 (.48)	1.80 (.47)
Discussed school courses with parents (1 = often; 0 = sometimes)	2.29 (.82)	1.33 (.39)	.97 (.59)	1.17 (.49)	.61 (.77)	.78 (.40)	1.63 (.36)*

Community/Volunteer/ Religious Variables							
Received a community service award (in first half of year) (1 = yes; 0 = no)	2.04 (1.24)	.96 (.57)	3.68 (.76)*	1.23 (.84)	9.21 (.81)**	1.00 (.61)	.46 (.70)
Among friends, how important to do community work/volunteer (1 = very; 0 = somewhat, not)	2.35 (.50)*	.54 (.34)*	.66 (.54)	.96 (.42)	.63 (.68)	1.29 (.32)	1.15 (.34)
How important to participate in religious activities (1 = very; 0 = somewhat, not)	.59 (.55)	2.07 (.39)*	1.48 (.58)	.45 (.45)*	.94 (.72)	1.58 (.37)*	1.24 (.63)
Peer Variables							
Among friends, how important get good grades (1 = very; 0 = somewhat, not)	.51 (.51)*	1.33 (.35)	1.55 (.55)	.73 (.56)	1.67 (.68)	.77 (.34)	.79 (.33)
Among friends, how important being popular (1 = very; 0 = somewhat or not)	.55 (.53)	.51 (.48)*	.55 (.67)	.56 (.43)*	.59 (.94)	.48 (.45)*	.65 (.40)
Important having strong friendships (1 = very; 0 = somewhat or not)	.66 (.49)	.99 (.34)	1.45 (.54)	1.23 (.47)	2.05 (.75)	1.78 (.34)*	2.14 (.34)**
SES	1.30 (.29)	.99 (.17)*	1.25 (.24)	1.13 (.21)	1.83 (.33)*	1.02 (.16)	.80 (.16)*
Constant	10.75 (1.00)**	.17 (.68)**	.07 (1.02)**	.05 (1.10)	.00 (2.02)**	.06 (.66)**	.17 (.63)**
Model χ^2 (df)	40.36 (13)**	16.35 (13)	25.47 (13)**	17.68 (13)*	42.51 (13)**	18.01 (13)*	21.84 (13)*

* Significant at .20 level.
** Significant at .05 level.
† Table shows antilogs and standard errors in parentheses.

TABLE A.4.5 Multiple Classification Analysis Showing Means for Young African American Women on Science Outcomes in Models Including Family, Community, Peer, and School Variables: Knowledge Networks Data†

	Expect Science Occupation at Age 30	Hope for Science Occupation at Age 30	High School Grades Science A's and B's	Good in Science	Like Science	Feel Welcome in Science
Family Variables						
How many children do you hope to have						
0 = 0 to 2	.28	.32	.42*	.27	.38	.63
1 = 3+	.23	.25	.32*	.26	.20	.65
Community Variables						
Important to give back to your community						
0 = not to somewhat	.29	.30	.32**	.23*	.28	.60*
1 = very important	.23	.29	.46**	.31*	.34	.69*
Peer Variables						
Students at your school tend to encourage you in science						
0 = no	.29**	.32**	.38	.25*	.29	.61*
1 = yes	.11**	.15**	.38	.37*	.40	.73*
School Variables						
Type of school						
0 = public	.24**	.27**	.42**	.29**	.33	.64
1 = private	.51**	.52**	.07**	.04**	.11	.61
An all-girls school						
0 = no	.28**	.31*	.39	.27	.32	.64
1 = yes	.09**	.01*	.23	.17	.09	.49

Percent minority students at your school						
0 = under 50%	.37**	.38**	.35	.31	.32	.72**
1 = 50% or over	.21**	.25**	.40	.25	.30	.59**
Adults at your school tend to encourage you in science						
0 = no	.28	.30	.30**	.21*	.25	.52**
1 = yes	.24	.29	.45**	.31*	.36	.73**
Other Variables						
Interest and abilities in science influenced by feedback from teachers, students, and family						
0 = no	.23*	.26*	.34*	.21**	.21	.56**
1 = yes	.32*	.36*	.45*	.37**	.48	.76**
F	2.91**	1.74*	3.98**	4.32**	4.35**	5.04**

* Significant at .20 level.
** Significant at .05 level.
† Household income and age are controlled in all equations.

TABLE A.5.1 Means (and Standard Deviations) on Family Variables for African American and White Women: NELS†

Family Variables	Race	
	African American	White
Mother's education (1 = college; 0 = less)	.14 (.35)*	.27 (.45)*
Father's education (1 = college; 0 = less)	.18 (.39)*	.35 (.48)*
Mother's occupation (1 = professional; 0 = other)	.12 (.32)	.13 (.34)
Father's occupation (1 = professional; 0 = other)	.10 (.31)*	.28 (.45)*
Mother's educational expectations for R (1 = college degree or more; 0 = other)	.87 (.34)*	.80 (.40)*
Father's educational expectations for R (1 = college degree or more; 0 = other)	.86 (.34)*	.80 (.40)*
Mothers's desire for R after high school (1 = college; 0 = not)	.85 (.36)*	.76 (.43)*
Father's desire for R after high school (1 = college; 0 = not)	.77 (.42)	.75 (.43)
Close relatives' desire for R after high school (1 = college; 0 = not)	.84 (.37)*	.77 (.42)*
Discussed school courses with parents (1 = often; 0 = sometimes or not)	.20 (.40)	.20 (.40)
Discussed grades with parents (1 = often; 0 = sometimes or not)	.52 (.50)*	.47 (.50)*
Importance of parents in science decisions (1 = very; 0 = somewhat or not)	.37 (.48)*	.25 (.43)*
Importance of siblings in science decisions (1 = very; 0 = somewhat or not)	.21 (.41)*	.08 (.27)*

* T-test for differences in means significant at .05 level.
† R = respondent.

TABLE A.5.2 Means (and Standard Deviations) on Science Outcomes for African American Women by Family Characteristics: NELS†

Family Variables	1988 (8th Grade) In Advanced, Enriched, or Accelerated Science Course	1990 (10th Grade) Coursework in Chemistry	Access 1992 (12th Grade) Enrolled in Science Classes in the Last Two Years	1994 Science Major at First Postsecondary Institution	2000 First Postsecondary Degree—Science	2000 Science Degree Expected by Age 30
Mother's education						
College	.35 (.48)	.25 (.44)*	.97 (.16)**	.35 (.48)*	.14 (.35)	.36 (.49)*
Less	.40 (.49)	.17 (.38)*	.88 (.33)**	.23 (.48)*	.15 (.36)	.26 (.44)*
Father's education						
College	.47 (.50)	.25 (.43)	.98 (.15)**	.33 (.48)*	.13 (.34)	.32 (.47)**
Less	.40 (.49)	.22 (.41)	.85 (.36)**	.24 (.43)*	.16 (.36)	.15 (.36)**
Mother's occupation						
Professional	.25 (.44)**	.16 (.37)	.85 (.36)	.30 (.46)	.12 (.32)	.16 (.37)*
Other	.41 (.49)**	.18 (.38)	.89 (.31)	.24 (.43)	.14 (.35)	.27 (.45)*
Father's occupation						
Professional	.30 (.46)	.31 (.46)*	.93 (.24)	.27 (.45)	.17 (.38)	.32 (.47)
Other	.39 (.49)	.19 (.39)*	.91 (.29)	.25 (.43)	.16 (.37)	.25 (.43)
Mother's educational expectations for R						
College+	.38 (.49)	.24 (.42)*	.90 (.31)	.23 (.42)	.15 (.36)	.18 (.39)
Other	.49 (.51)	.11 (.32)*	.95 (.22)	.18 (.39)	.07 (.27)	.25 (.44)
Father's educational expectations for R						
College+	.39 (.49)	.27 (.45)**	.95 (.23)	.25 (.43)	.18 (.39)*	.20 (.40)
Other	.42 (.50)	.09 (.29)**	.93 (.26)	.21 (.42)	.00 (.00)*	.27 (.46)

TABLE A.5.2 *Continued*

Family Variables	Access					
	1988 (8th Grade) In Advanced, Enriched, or Accelerated Science Course	1990 (10th Grade) Coursework in Chemistry	1992 (12th Grade) Enrolled in Science Classes in the Last Two Years	1994 Science Major at First Postsecondary Institution	2000 First Postsecondary Degree—Science	2000 Science Degree Expected by Age 30
Mother's desire for R after high school						
College	.38 (.49)*	.23 (.42)**	.89 (.31)	.25 (.43)	.14 (.35)	.18 (.38)
Not	.49 (.50)*	.08 (.28)**	.88 (.33)	.35 (.49)	.06 (.25)	.27 (.45)
Father's desire for R after high school						
College	.37 (.48)	.27 (.44)**	.95 (.23)*	.25 (.43)	.17 (.38)	.24 (.43)**
Not	.42 (.50)	.13 (.34)**	.90 (.30)*	.23 (.43)	.08 (.29)	.42 (.50)**
Close relatives' desire for R after high school						
College	.36 (.48)*	.17 (.38)**	.90 (.31)	.22 (.42)	.15 (.36)*	.24 (.43)*
Not	.51 (.50)*	.34 (.48)**	.84 (.37)	.26 (.45)	.06 (.24)*	.08 (.28)*
Discussed school courses with parents						
Often	.36 (.48)	.18 (.39)	.96 (.20)**	.32 (.47)*	.15 (.36)	.21 (.41)
Sometimes, Never	.38 (.49)	.17 (.38)	.86 (.34)**	.22 (.42)*	.13 (.34)	.25 (.43)
Discussed grades with parents						
Often	.34 (.48)	.24 (.43)*	.88 (.32)	.24 (.43)	.12 (.33)	.19 (.40)*
Sometimes, Never	.41 (.49)	.17 (.38)*	.89 (.30)	.22 (.42)	.14 (.35)	.29 (.46)*
Importance of parents in science decisions						
Very	.50 (.50)*	.20 (.41)	‡	.30 (.47)*	.27 (.45)	.30 (.47)
Some, Not	.37 (.49)*	.25 (.44)	‡	.44 (.50)*	.16 (.37)	.34 (.48)
Importance of siblings in science decisions						
Very	.32 (.48)	.78 (.43)**	‡	.22 (.44)*	.06 (.25)*	.17 (.40)*
Some, Not	.41 (.50)	.40 (.50)**	‡	.45 (.50)*	.29 (.46)*	.43 (.50)*

			Achievement			
Family Variables	1988 (8th Grade) Science Grades from 6th Grade Until Now (Mostly A's or A's & B's)	1988 (8th Grade) Science Standardized Score (Top Quartile)	1990 (10th Grade) Science Grades (Mostly A's or A's & B's)	1990 (10th Grade) Science Standardized Score (Top Quartile)	1992 (12th Grade) Science Standardized Score (Top Quartile)	2000 Current/Most Recent Occupation—Science
Mother's education						
College	.44 (.50)*	.27 (.45)**	.57 (.50)**	.17 (.38)**	.20 (.40)**	.24 (.43)
Less	.34 (.48)*	.08 (.27)**	.34 (.48)**	.05 (.22)**	.04 (.19)**	.24 (.43)
Father's education						
College	.50 (.50)**	.30 (.46)**	.45 (.50)**	.20 (.40)**	.22 (.40)**	.17 (.38)*
Less	.33 (.47)**	.07 (.26)**	.30 (.46)**	.06 (.23)**	.03 (.48)**	.25 (.43)*
Mother's occupation						
Professional	.26 (.45)	.28 (.46)**	.55 (.50)**	.08 (.27)	.16 (.37)**	.21 (.41)
Other	.35 (.48)	.08 (.27)**	.35 (.48)**	.06 (.24)	.05 (.22)**	.23 (.41)
Father's occupation						
Professional	.34 (.48)	.29 (.46)**	.29 (.46)*	.16 (.36)**	.17 (.38)**	.19 (.40)
Other	.35 (.48)	.08 (.27)**	.42 (.49)*	.05 (.21)**	.04 (.21)**	.25 (.43)
Mother's educational expectations for R						
College+	.38 (.49)**	.14 (.34)**	.37 (.48)	.09 (.28)*	.09 (.28)**	.22 (.42)
Other	.19 (.39)**	.03 (.17)**	.41 (.49)	.02 (.13)*	.00 (.00)**	.16 (.37)
Father's educational expectations for R						
College+	.38 (.40)*	.15 (.36)*	.41 (.49)*	.09 (.28)	.08 (.27)	.28 (.44)
Other	.25 (.44)*	.04 (.20)*	.27 (.45)*	.04 (.20)	.02 (.15)	.19 (.40)
Mother's desire for R after high school						
College	.35 (.48)**	.13 (.34)**	.37 (.48)	.08 (.27)*	.07 (.25)	.24 (.43)*
Not	.17 (.38)**	.02 (.15)**	.36 (.48)	.02 (.14)*	.05 (.22)	.13 (.34)*

TABLE A.5.2 *Continued*

	Achievement					
Family Variables	1988 (8th Grade) Science Grades from 6th Grade Until Now (Mostly A's or A's & B's)	1988 (8th Grade) Science Standardized Score (Top Quartile)	1990 (10th Grade) Science Grades (Mostly A's or A's & B's)	1990 (10th Grade) Science Standardized Score (Top Quartile)	1992 (12th Grade) Science Standardized Score (Top Quartile)	2000 Current/Most Recent Occupation—Science
Father's desire for R after high school						
College	.38 (.49)**	.14 (.35)**	.38 (.49)	.08 (.28)*	.07 (.26)*	.28 (.45)
Not	.09 (.29)**	.04 (.19)**	.40 (.49)	.03 (.18)*	.02 (.15)*	.32 (.47)
Close relatives' desire for R after high school						
College	.31 (.46)	.12 (.33)*	.41 (.49)**	.08 (.26)	.07 (.26)	.23 (.42)*
Not	.31 (.46)	.04 (.19)*	.22 (.42)**	.04 (.19)	.03 (.16)	.33 (.47)*
Discussed school courses with parents						
Often	.34 (.47)	.16 (.37)*	.34 (.48)	.09 (.29)	.13 (.34)**	.22 (.41)
Sometimes, Never	.29 (.46)	.10 (.30)*	.42 (.49)	.07 (.26)	.06 (.24)**	.22 (.41)
Discussed grades with parents						
Often	.38 (.49)**	.10 (.30)	.35 (.48)*	.05 (.23)*	.08 (.27)	.24 (.43)
Sometimes, Never	.28 (.45)**	.13 (.33)	.43 (.49)*	.09 (.29)*	.06 (.24)	.26 (.44)
Importance of parents in science decisions						
Very	.30 (.47)	.09 (.29)**	.47 (.51)	.17 (.38)	.12 (.33)	.34 (.48)*
Some, Not	.40 (.49)	.28 (.45)**	.50 (.50)	.13 (.34)	.16 (.37)	.21 (.41)*
Importance of siblings in science decisions						
Very	.11 (.32)**	.00 (.00)**	.31 (.48)*	.00 (.00)*	.00 (.00)**	.45 (.51)**
Some, Not	.40 (.49)**	.27 (.45)**	.58 (.50)*	.16 (.36)*	.19 (.40)**	.21 (.41)**

				Attitudes			
Family Variables	1988 (8th Grade) Usually Look Forward to Science Class	1988 (8th Grade) Science Will Be Useful in My Future	1990 (10th Grade) Often Work Hard in Science Class	1992 (12th Grade) Interested in Science	1992 (12th Grade) Does Well in Science	1992 (12th Grade) Need Science for Job After High School	2000 Planned Occupation by Age 30— Science
Mother's education							
College	.66 (.48)	.81 (.39)*	.62 (.49)	.45 (.50)	.55 (.51)	.25 (.44)	.17 (.38)**
Less	.60 (.49)	.73 (.45)*	.68 (.47)	.45 (.50)	.46 (.50)	.34 (.47)	.39 (.49)**
Father's education							
College	.73 (.45)*	.79 (.41)	.64 (.48)	.49 (.51)	.58 (.50)*	.35 (.48)	.34 (.49)
Less	.62 (.49)*	.73 (.45)	.64 (.48)	.43 (.50)	.44 (.50)*	.34 (.47)	.35 (.48)
Mother's occupation							
Professional	.58 (.49)	.62 (.49)*	.57 (.50)	.58 (.51)*	.65 (.49)*	.27 (.46)	.21 (.41)*
Other	.59 (.49)	.71 (.45)*	.65 (.48)	.39 (.49)*	.45 (.50)*	.30 (.46)	.32 (.47)*
Father's occupation							
Professional	.62 (.49)	.72 (.45)	.58 (.50)	.57 (.50)*	.63 (.49)*	.32 (.48)	.35 (.48)
Other	.58 (.49)	.69 (.46)	.65 (.48)	.41 (.49)*	.46 (.50)*	.29 (.46)	.27 (.44)
Mother's educational expectations for R							
College+	.63 (.48)	.77 (.42)**	.64 (.48)*	.48 (.50)*	.51 (.50)	.35 (.48)	.29 (.45)
Other	.59 (.49)	.64 (.49)**	.53 (.51)*	.22 (.44)*	.38 (.51)	.11 (.34)	.28 (.45)
Father's educational expectations for R							
College+	.65 (.48)	.75 (.44)	.65 (.48)*	.48 (.50)**	.51 (.50)	.33 (.47)	.28 (.45)
Other	.65 (.48)	.65 (.48)	.52 (.50)*	.43 (.35)**	.34 (.50)	.24 (.47)	.28 (.46)

TABLE A.5.2 *Continued*

				Attitudes			
Family Variables	1988 (8th Grade) Usually Look Forward to Science Class	1988 (8th Grade) Science Will Be Useful in My Future	1990 (10th Grade) Often Work Hard in Science Class	1992 (12th Grade) Interested in Science	1992 (12th Grade) Does Well in Science	1992 (12th Grade) Need Science for Job After High School	2000 Planned Occupation by Age 30—Science
Mother's desire for R after high school							
College	.62 (.49)	.72 (.45)*	.58 (.49)*	.46 (.50)**	.52 (.50)*	.35 (.48)*	.29 (.46)
Not	.61 (.49)	.63 (.48)*	.69 (.47)*	.20 (.42)**	.33 (.49)*	.15 (.37)*	.25 (.44)
Father's desire for R after high school							
College	.68 (.47)**	.71 (.45)**	.57 (.49)**	.45 (.50)*	.49 (.50)	.31 (.47)	.28 (.45)*
Not	.50 (.50)**	.59 (.50)**	.85 (.36)**	.25 (.25)*	.44 (.51)	.17 (.40)	.40 (.49)*
Close relatives' desire for R after high school							
College	.58 (.49)**	.68 (.47)	.62 (.49)	.46 (.50)*	.50 (.05)	.34 (.00)**	.34 (.47)*
Not	.75 (.43)**	.66 (.48)	.55 (.50)	.29 (.47)*	.51 (.12)	.14 (.35)**	.00 (.00)*
Discussed school courses with parents							
Often	.71 (.46)**	.76 (.43)*	.81 (.40)**	.40 (.50)	.50 (.50)	.32 (.47)	.38 (.49)
Sometimes, Never	.54 (.50)**	.66 (.48)*	.57 (.50)**	.42 (.50)	.47 (.50)	.28 (.45)	.33 (.33)
Discussed grades with parents							
Often	.61 (.49)*	.72 (.45)*	.59 (.49)	.51 (.50)*	.66 (.48)**	.34 (.48)	.33 (.47)
Sometimes, Never	.54 (.50)*	.64 (.48)*	.60 (.49)	.34 (.48)*	.34 (.48)**	.26 (.44)	.33 (.47)

Importance of parents in science decisions							
Very	.69 (.47)	.70 (.46)	.68 (.47)	.34 (.47)*	.50 (.50)	.68 (.47)	.27 (.45)
Some, Not	.60 (.49)	.75 (.43)	.60 (.49)	.49 (.50)*	.48 (.50)	.73 (.44)	.37 (.49)
Importance of siblings in science decisions							
Very	.45 (.51)**	.34 (.49)**	.46 (.51)	.57 (.51)	.40 (.52)	.42 (.52)*	.54 (.52)**
Some, Not	.63 (.41)**	.83 (.38)**	.55 (.50)	.43 (.50)	.50 (.50)	.19 (.40)*	.22 (.42)**

* T-test for difference in means significant at .20 level.
** T-test for difference in means significant at .05 level.
† R = respondent.
‡ In some situations the difference in means test was not possible due to small N.

TABLE A.5.3 Results from Logistic Regression Models Showing Effects of Family Variables on Science Outcomes for African American Women: NELS†

Family Variables	Access				Achievement		Attitudes
	1992 (12th Grade) Enrolled in Science Classes in the Last Two Years	1994 Science Major at First Postsecondary Institution	2000 First Postsecondary Degree—Science	2000 Science Degree Expected by Age 30	1992 (12th Grade) Science Standardized Score (Top Quartile)	2000 Current/Most Recent Occupation—Science	2000 Planned Occupation by Age 30—Science
Mother's education (1 = college; 0 = less)	‡	3.83 (.66)**	.46 (.99)	1.84 (.84)	2.29 (.87)	.98 (.63)	.68 (.64)
Father's education (1 = college; 0 = less)	‡	.63 (.61)	.25 (.97)*	7.17 (.88)**	1.39 (.80)	.40 (.63)*	.97 (.57)
Mother's occupation (1 = professional; 0 = other)	.15 (1.34)*	.80 (.73)	.41 (1.12)	.21 (1.49)	2.31 (.90)	2.42 (.65)*	1.99 (.71)
Father's occupation (1 = professional; 0 = other)	‡	.46 (.67)	.77 (.97)	2.29 (.77)	.69 (.92)	1.39 (.61)	2.80 (.59)*
Mother's educational expectations for R (1 = college degree or more; 0 = other)	.55 (1.59)	.72 (1.19)	‡	.51 (1.82)	‡	.78 (1.32)	.57 (1.34)
Father's educational expectations for R (1 = college degree or more; 0 = other)	1.34 (1.37)	1.61 (1.06)	‡	.25 (1.74)	.20 (1.65)	3.03 (1.26)	5.64 (1.31)*
Mother's desire for R after high school (1 = college; 0 = not)	1.37 (2.22)	1.68 (1.17)	‡	.10 (1.87)	‡	1.31 (.99)	1.24 (1.09)

Father's desire for R after high school (1 = college; 0 = not)	1.95 (1.49)	.59 (1.08)	‡	24.47 (2.45)*	.62 (1.89)	.29 (.92)*	.26 (1.09)
Close relatives' desire for R after high school (1 = college; 0 = not)	.27 (1.69)	.67 (1.04)	.38 (1.78)	.20 (1.99)	.54 (1.80)	2.59 (.89)	1.39 (.92)
Discussed school courses with parents (1 = often; 0 = sometimes)	2.43 (1.44)	3.76 (.51)**	1.52 (.69)	.88 (.78)	.77 (.85)	1.05 (.47)	1.76 (.47)
SES	2.99 (.61)*	.96 (.87)	1.67 (.37)*	.54 (.36)*	2.68 (.48)**	1.19 (.24)	.76 (.23)
Constant	2.76 (1.95)	.43 (1.08)	.00 (‡)	4.59 (1.88)	.00 (‡)	.10 (.99)**	.53 (1.13)
Model χ^2 (df)	14.30 (11)	12.59 (11)	14.92 (11)*	20.34 (11)**	21.14 (11)**	9.17 (11)	12.80 (11)

* Significant at .20 level.

** Significant at .05 level.

† Table shows antilogs and standard errors in parentheses. R = respondent.

‡ Number is very large or small.

TABLE A.5.4   Means (and Standard Deviations) on Community/Volunteer/
Religious Variables for African American and White Women: NELS†

Community/Volunteer/Religious Variables	Race	
	African American	White
How often do non–school-sponsored volunteer or community service? (1 = once a week or more; 0 = less than once a week)	.13 (.33)	.12 (.33)
During past 2 years performed unpaid volunteer or community service work? (1 = yes; 0 = no)	.43 (.50)*	.55 (.50)*
Among friends, how important to do community work/volunteer (1 = very; 0 = somewhat, not)	.08 (.27)*	.03 (.17)*
R volunteered with youth organizations in last 12 months? (1 = yes; 0 = no)	.13 (.34)*	.21 (.41)*
R volunteered with civic or community organization in last 12 months? (1 = yes; 0 = no)	.27 (.45)	.25 (.43)
R volunteered with church group in last 12 months? (1 = yes; 0 = no)	.52 (.50)	.53 (.50)
Received a community service award (in first half of year) (1 = yes; 0 = no)	.08 (.28)	.07 (.25)
R's volunteer work strongly encouraged by someone else (1 = yes; 0 = no)	.38 (.49)	.40 (.49)
How often R attends religious activities (1 = once a week or more; 0 = less than once a week)	.44 (.50)*	.37 (.48)*
Among friends, how important to participate in religious activities (1 = very; 0 = somewhat, not)	.16 (.37)*	.11 (.31)*

*T-test is significant at .05 level.
†R = respondent.

TABLE A.5.5 Means (and Standard Deviations) on Science Outcomes for African American Women by Community/Volunteer/Religious Characteristics: NELS†

Community/Volunteer/Religious Variables	1988 (8th Grade) In Advanced, Enriched, or Accelerated Science Course	1990 (10th Grade) Coursework in Chemistry	Access 1992 (12th Grade) Enrolled in Science Classes in the Last Two Years	1994 Science Major at First Postsecondary Institution	2000 First Postsecondary Degree—Science	2000 Science Degree Expected by Age 30
How often do non-school-sponsored volunteer or community service?						
Once a week or more	.29 (.46)	.10 (.31)*	.93 (.26)	.29 (.46)	.10 (.31)	.31 (.47)
Less than once a week	.38 (.49)	.20 (.40)*	.89 (.31)	.20 (.40)	.13 (.34)	.19 (.40)
During past 2 years performed unpaid volunteer or community service work?						
Yes	.41 (.49)*	.18 (.38)	.95 (.38)**	.24 (.43)	.16 (.37)	.23 (.42)
No	.33 (.47)*	.20 (.40)	.84 (.40)**	.20 (.40)	.11 (.31)	.21 (.41)
Among friends, how important to do community work/volunteer						
Very	.38 (.49)	.22 (.42)*	.93 (.25)**	.17 (.37)**	.11 (.31)	.27 (.45)
Somewhat, Not	.38 (.49)	.17 (.37)*	.84 (.37)**	.29 (.54)**	.14 (.35)	.21 (.41)
R volunteered with youth organizations in last 12 months?						
Yes	.56 (.50)*	.22 (.42)	.90 (.31)*	.37 (.49)*	.14 (.36)	.27 (.46)*
No	.39 (.49)*	.17 (.38)	.96 (.20)*	.23 (.42)*	.25 (.43)	.14 (.35)*
R volunteered with civic or community organization in last 12 months?						
Yes	.29 (.46)*	.20 (.41)	.96 (.20)	.18 (.39)	.12 (.33)	.17 (.38)
No	.45 (.50)*	.17 (.38)	.94 (.23)	.27 (.45)	.19 (.39)	.26 (.44)

TABLE A.5.5 *Continued*

			Access			
Community/Volunteer/Religious Variables	1988 (8th Grade) In Advanced, Enriched, or Accelerated Science Course	1990 (10th Grade) Coursework in Chemistry	1992 (12th Grade) Enrolled in Science Classes in the Last Two Years	1994 Science Major at First Postsecondary Institution	2000 First Postsecondary Degree—Science	2000 Science Degree Expected by Age 30
R volunteered with church group in last 12 months?						
Yes	.52 (.50)**	.19 (.39)	.94 (.24)	.22 (.42)	.15 (.35)	.16 (.37)**
No	.30 (.46)**	.17 (.38)	.96 (.20)	.27 (.45)	.15 (.36)	.33 (.48)**
Received a community service award (in first half of year)						
Yes	.67 (.48)**	.25 (.44)	.98 (.15)*	.33 (.48)	.28 (.46)**	.11 (.33)*
No	.36 (.48)**	.19 (.40)	.88 (.33)*	.23 (.42)	.13 (.33)**	.25 (.43)*
R's volunteer work strongly encouraged by someone else						
Yes	.44 (.50)	.20 (.41)	.97 (.18)	.23 (.42)	.15 (.36)	.15 (.36)**
No	.39 (.49)	.16 (.37)	.94 (.24)	.25 (.44)	.17 (.38)	.29 (.46)**
How often R attends religious activities						
Once a week or more	.45 (.50)**	.26 (.44)**	.94 (.25)**	.24 (.43)	.15 (.36)	.14 (.35)**
Less than once a week	.30 (.46)**	.13 (.34)**	.85 (.35)**	.21 (.41)	.12 (.32)	.29 (.46)**
Among friends, how important to participate in religious activities						
Very	.38 (.49)	.23 (.42)**	.91 (.29)**	.24 (.43)	.16 (.36)	.24 (.43)
Somewhat, Not	.38 (.49)	.12 (.33)**	.85 (.36)**	.19 (.40)	.08 (.27)	.25 (.44)

	Achievement					
Community/Volunteer/Religious Variables	1988 (8th Grade) Science Grades from 6th Grade Until Now (Mostly A's or A's & B's)	1988 (8th Grade) Science Standardized Score (Top Quartile)	1990 (10th Grade) Science Grades (Mostly A's or A's & B's)	1990 (10th Grade) Science Standardized Score (Top Quartile)	1992 (12th Grade) Science Standardized Score (Top Quartile)	2000 Current/Most Recent Occupation—Science
How often do non-school-sponsored volunteer or community service?						
Once a week or more	.19 (.40)**	.18 (.39)**	.43 (.50)	.06 (.23)	.08 (.28)	.16 (.37)*
Less than once a week	.33 (.47)**	.09 (.29)**	.40 (.49)	.07 (.25)	.06 (.24)	.25 (.43)*
During past 2 years performed unpaid volunteer or community service work?						
Yes	.37 (.48)**	.18 (.39)**	.43 (.50)*	.11 (.31)**	.11 (.31)**	.24 (.43)
No	.27 (.44)**	.04 (.20)**	.36 (.48)*	.03 (.18)**	.02 (.15)**	.23 (.42)
Among friends, how important to do community work/volunteer						
Very	.34 (.47)	.07 (.26)**	.41 (.49)*	.07 (.25)	.05 (.21)*	.33 (.47)**
Somewhat, Not	.30 (.46)	.14 (.35)**	.34 (.48)*	.08 (.27)	.08 (.28)*	.16 (.37)**
R volunteered with youth organizations in last 12 months?						
Yes	.31 (.47)	.27 (.45)	.32 (.48)	.17 (.39)	.16 (.37)	.30 (.47)
No	.37 (.48)	.18 (.38)	.45 (.49)	.10 (.30)	.11 (.31)	.24 (.43)
R volunteered with civic or community organization in last 12 months?						
Yes	.50 (.51)**	.21 (.41)	.60 (.50)**	.11 (.32)	.17 (.38)*	.16 (.37)*
No	.31 (.47)**	.19 (.39)	.39 (.49)**	.11 (.32)	.08 (.27)*	.28 (.45)*

TABLE A.5.5 *Continued*

	Achievement					
Community/Volunteer/Religious Variables	1988 (8th Grade) Science Grades from 6th Grade Until Now (Mostly A's & B's)	1988 (8th Grade) Science Standardized Score (Top Quartile)	1990 (10th Grade) Science Grades (Mostly A's or A's & B's)	1990 (10th Grade) Science Standardized Score (Top Quartile)	1992 (12th Grade) Science Standardized Score (Top Quartile)	2000 Current/Most Recent Occupation—Science
R volunteered with church group in last 12 months?						
Yes	.34 (.48)	.11 (.32)**	.34 (.48)**	.14 (.34)*	.09 (.28)	.25 (.44)
No	.41 (.50)	.28 (.45)**	.52 (.50)**	.08 (.27)*	.14 (.35)	.21 (.41)
Received a community service award (in first half of year)						
Yes	.31 (.47)	.19 (.40)*	.58 (.50)**	.19 (.40)**	.25 (.40)**	.16 (.37)
No	.31 (.46)	.10 (.30)*	.34 (.48)**	.06 (.23)**	.04 (.20)**	.25 (.43)
R's volunteer work strongly encouraged by someone else						
Yes	.32 (.47)	.21 (.41)	.40 (.49)	.20 (.40)**	.13 (.33)	.32 (.47)**
No	.40 (.49)	.17 (.38)	.45 (.50)	.05 (.22)**	.10 (.30)	.19 (.39)**
How often R attends religious activities						
Once a week or more	.39 (.49)**	.13 (.34)*	.37 (.48)	.09 (.29)*	.07 (.25)	.28 (.45)**
Less than once a week	.25 (.43)**	.08 (.27)*	.41 (.49)	.04 (.21)*	.05 (.23)	.19 (.40)**
Among friends, how important to participate in religious activities						
Very	.35 (.48)*	.10 (.30)	.40 (.49)*	.08 (.27)	.06 (.23)	.29 (.46)**
Somewhat, Not	.26 (.44)*	.12 (.33)	.33 (.47)*	.06 (.24)	.08 (.28)	.17 (.37)**

		Attitudes					
Community/Volunteer/Religious Variables	1988 (8th Grade) Usually Look Forward to Science Class	1988 (8th Grade) Science Will Be Useful in My Future	1990 (10th Grade) Often Work Hard in Science Class	1992 (12th Grade) Interested in Science	1992 (12th Grade) Does Well in Science	1992 (12th Grade) Need Science for Job After High School	2000 Planned Occupation by Age 30—Science
How often do non-school-sponsored volunteer or community service?							
Once a week or more	.56 (.50)	.68 (.47)	.84 (.37)**	.43 (.51)	.51 (.51)	.11 (.32)**	.33 (.48)
Less than once a week	.57 (.50)	.68 (.47)	.59 (.49)**	.42 (.50)	.45 (.50)	.36 (.48)**	.30 (.46)
During past 2 years performed unpaid volunteer or community service work?							
Yes	.63 (.48)**	.66 (.48)	.63 (.49)	.48 (.50)*	.51 (.50)	.38 (.49)**	.31 (.46)
No	.52 (.50)**	.70 (.46)	.63 (.48)	.33 (.48)*	.44 (.50)	.19 (.40)**	.30 (.46)
Among friends, how important to do community work/volunteer							
Very	.59 (.49)	.61 (.49)**	.58 (.50)	.36 (.49)	.46 (.50)	.28 (.45)	.33 (.47)
Somewhat, Not	.57 (.50)	.73 (.45)**	.63 (.48)	.47 (.50)	.47 (.50)	.31 (.47)	.33 (.47)
R volunteered with youth organizations in last 12 months?							
Yes	.72 (.46)	.80 (.41)*	.85 (.37)**	.62 (.51)	.64 (.50)	.72 (.48)**	.32 (.48)
No	.62 (.49)	.64 (.48)*	.59 (.49)**	.47 (.50)	.49 (.50)	.34 (.48)**	.31 (.46)
R volunteered with civic or community organization in last 12 months?							
Yes	.51 (.51)**	.54 (.50)**	.43 (.50)**	.36 (.50)	.41 (.51)	.25 (.45)	.20 (.40)**
No	.67 (.47)**	.70 (.46)**	.68 (.47)**	.52 (.50)	.53 (.50)	.42 (.50)	.36 (.48)**

TABLE A.5.5 *Continued*

				Attitudes			
Community/Volunteer/Religious Variables	1988 (8th Grade) Usually Look Forward to Science Class	1988 (8th Grade) Science Will Be Useful in My Future	1990 (10th Grade) Often Work Hard in Science Class	1992 (12th Grade) Interested in Science	1992 (12th Grade) Does Well in Science	1992 (12th Grade) Need Science for Job After High School	2000 Planned Occupation by Age 30— Science
R volunteered with church group in last 12 months?							
Yes	.71 (.46)**	.68 (.47)	.66 (.48)	.45 (.50)	.45 (.50)	.31 (.47)	.48 (.51)*
No	.56 (.50)**	.62 (.49)	.61 (.49)	.53 (.51)	.53 (.51)	.31 (.47)	.30 (.47)*
Received a community service award (in first half of year)							
Yes	.75 (.44)*	.91 (.30)**	.74 (.45)*	.58 (.52)	.52 (.52)	.00 (.00)*	.32 (.48)
No	.58 (.50)*	.66 (.48)**	.60 (.49)*	.40 (.49)	.48 (.50)	.32 (.47)*	.33 (.47)
R's volunteer work strongly encouraged by someone else							
Yes	.62 (.49)	.68 (.47)	.61 (.49)	.45 (.50)	.36 (.48)**	.46 (.51)	.37 (.49)*
No	.64 (.48)	.65 (.48)	.64 (.48)	.51 (.51)	.64 (.49)**	.31 (.47)	.27 (.45)*
How often R attends religious activities							
Once a week or more	.65 (.48)**	.68 (.47)	.58 (.50)*	.47 (.50)*	.52 (.50)	.42 (.50)*	.23 (.42)**
Less than once a week	.51 (.50)**	.68 (.47)	.67 (.47)*	.36 (.48)*	.43 (.50)	.36 (.48)*	.36 (.48)**
Among friends, how important to participate in religious activities							
Very	.63 (.49)**	.69 (.47)	.63 (.48)*	.42 (.50)	.46 (.50)	.27 (.45)	.32 (.47)
Somewhat, Not	.50 (.50)**	.64 (.48)	.54 (.50)*	.43 (.50)	.47 (.51)	.34 (.48)	.34 (.48)

* T-Test for difference in means significant at .20 level.

** T-Test for difference in means significant at .05 level.

† R = respondent.

TABLE A.5.6 Results from Logistic Regression Models Showing Effects of Community/Volunteer/Religious Variables on Science Outcomes for African American Women: NELS†

Community/Volunteer/Religious Variables	Access				Achievement		Attitudes
	1992 (12th Grade) Enrolled in Science Classes in the Last Two Years	1994 Science Major at First Postsecondary Institution	2000 First Postsecondary Degree—Science	2000 Science Degree Expected by Age 30	1992 (12th Grade) Science Standardized Score (Top Quartile)	2000 Current/Most Recent Occupation—Science	2000 Planned Occupation by Age 30—Science
How often do non-school-sponsored volunteer or community service? (1 = once a week or more; 0 = less than once a week)	.72 (.62)	1.39 (.43)	.68 (.75)	1.76 (.47)	.30 (.74)*	.49 (.46)*	1.15 (.38)
During past 2 years performed unpaid volunteer or community service work? (1 = yes; 0 = no)	2.86 (.46)**	1.19 (.34)	.96 (.53)	1.55 (.42)	3.64 (.63)**	.88 (.28)	1.31 (.28)
Among friends, how important to do community work/volunteer (1 = very; 0 = somewhat, not)	3.53 (.43)**	.43 (.32)**	.29 (.53)**	1.81 (.37)*	.72 (.60)	1.90 (.28)**	1.09 (.27)
Received a community service award (in first half of year) (1 = yes; 0 = no)	2.87 (1.19)	1.20 (.52)	2.58 (.63)*	.18 (1.34)	11.30 (.69)**	.40 (.63)*	.55 (.64)

TABLE A.5.6 *Continued*

Community/Volunteer/Religious Variables	Access				Achievement		Attitudes
	1992 (12th Grade) Enrolled in Science Classes in the Last Two Years	1994 Science Major at First Postsecondary Institution	2000 First Postsecondary Degree—Science	2000 Science Degree Expected by Age 30	1992 (12th Grade) Science Standardized Score (Top Quartile)	2000 Current/Most Recent Occupation—Science	2000 Planned Occupation by Age 30—Science
How often R attends religious activities (1 = once a week or more; 0 = less than once a week)	1.46 (.45)	.79 (.35)	1.13 (.55)	.29 (.42)**	.67 (.62)	1.26 (.29)	.53 (.29)**
How important to participate in religious activities (1 = very; 0 = somewhat, not)	1.23 (.40)	1.72 (.37)**	2.81 (.56)*	.97 (.39)	.75 (.67)	1.86 (.32)**	1.26 (.29)
SES	1.79 (.21)**	1.06 (.14)	1.20 (.20)	.82 (.18)	2.77 (.28)**	1.22 (.12)*	.69 (.13)**
Constant	.95 (.42)	.25 (.46)**	.06 (.66)**	.45 (.51)*	.00 (.96)**	.11 (.38)*	1.02 (.33)
Model χ^2 (df)	41.31 (7)**	10.21 (7)*	11.04 (7)*	22.50 (7)**	48.46 (7)**	27.67 (7)**	19.31 (7)**

* Significant at .20 level.

** Significant at .05 level.

† Table shows antilogs and standard errors in parentheses. R = respondent.

TABLE A.5.7 OLS and Logistic† Regression Models Using Family and Control Variables to Predict Science Outcomes for Young African American Women: Knowledge Networks Data‡

	Science Outcomes					
	OLS Models			Logistic Models		
Family Variables	High School Science Grades	Good in Science	Likes Science	Expect Science Occupation at Age 30	Hope for Science Occupation at Age 30	Feel Welcome in Science
Mother has a degree in science (1 = yes; 0 = no)	.86 (.46)*	.85 (.38)**	-.03 (.45)	.76 (.95)	1.98 (.84)	2.80 (1.04)
Father has a degree in science (1 = yes; 0 = no)	-.58 (.60)	-.74 (.50)*	-.95 (.59)*	.21 (1.89)	.08 (1.76)*	.30 (.99)
Mother has a job in science (1 = yes; 0 = no)	.21 (.58)	-.31 (.48)	.43 (.57)	.14 (1.66)	.61 (1.09)	1.26 (1.26)
Father has a job in science (1 = yes; 0 = no)	.27 (1.07)	1.34 (.90)*	1.57 (1.05)*	5.64 (2.21)	4.77 (2.24)	.89 (1.75)
Mother's educational aspirations for R (1 = college degree or more; 0 = other)	-.43 (.31)*	-.29 (.26)	.13 (.31)	1.82 (.62)	.95 (.58)	1.36 (.51)
Father's educational aspirations for R (1 = college degree or more; 0 = other)	.49 (.31)**	.22 (.26)	-.22 (.31)	.54 (.59)	1.54 (.56)	2.42 (.51)*
Closeness to mother (5 = very; 1 = not at all)	-.14 (.08)*	.10 (.07)*	.00 (.08)	1.37 (.17)*	1.62 (.17)**	.98 (.14)
Closeness to father (5 = very; 1 = not at all)	.14 (.07)*	.01 (.06)	.08 (.07)	.82 (.15)*	1.02 (.14)	1.23 (.13)*
Mother's involvement in R's school (5 = considerable; 1 = none)	.00 (.08)	-.07 (.06)	-.12 (.07)*	1.00 (.14)	.89 (.14)	.94 (.13)

TABLE A.5.7 *Continued*

	Science Outcomes					
	OLS Models			Logistic Models		
Family Variables	High School Science Grades	Good in Science	Likes Science	Expect Science Occupation at Age 30	Hope for Science Occupation at Age 30	Feel Welcome in Science
Father's involvement in R's school (5 = considerable; 1 = none)	–.04 (.08)	–.12 (.07)*	–.21 (.08)**	1.09 (.16)	.98 (.15)	.91 (.14)
Importance of family in future (5 = very important; 1 = not at all important)	.04 (.09)	–.14 (.07)*	–.14 (.09)*	.99 (.16)	.75 (.15)**	.88 (.16)
Importance of work in future (5 = very important; 1 = not at all important)	.18 (.12)*	.24 (.10)**	.30 (.12)**	1.05 (.24)	.92 (.22)	1.34 (.21)*
Family encouragement in science (1 = yes; 0 = no)	.32 (.19)*	.53 (.16)**	.65 (.18)**	3.05 (.34)**	2.55 (.34)**	1.87 (.32)**
Constant	2.38 (.79)**	2.25 (.66)**	2.21 (.78)**	.03 (1.64)**	.07 (1.53)*	.17 (1.34)*
F (df)	1.82 (16)**	2.84 (16)**	2.29 (16)**			
R^2 (adjusted)	.05	.10	.07			
Model χ^2 (df)				34.95 (16)**	40.66 (16)**	32.26 (16)**

* Significant at .20 level.

** Significant at .05 level.

† Coefficients are presented as antilogs.

‡ Control variables include age, family income, and rural–urban residence. R = respondent.

TABLE A.6.1 Means (and Standard Deviations) on Peer Variables for African American and White Women: NELS†

	Race	
Peer Variables	African American	White
Importance of having strong friendships (1 = very; 0 = somewhat or not)	.61 (.49)*	.86 (.35)*
Among friends, how important to study (1 = very; 0 = somewhat or not)	.46 (.50)*	.43 (.50)*
Among friends, how important to get good grades (1 = very; 0 = somewhat or not)	.63 (.48)*	.55 (.50)*
Among friends, how important to finish high school (1 = very; 0 = somewhat or not)	.82 (.39)*	.89 (.31)*
Among friends, how important being popular (1 = very; 0 = somewhat or not)	.20 (.40)*	.24 (.43)*
Among friends, how important to have sexual relations (1 = very; 0 = somewhat or not)	.08 (.27)*	.10 (.30)*
Importance of friends in science decision (1 = very; 0 = somewhat or not	.19 (.39)	.20 (.40)
Friends' desire for R after high school (1 = college; 0 = not)	.64 (.48)*	.57 (.50)*
R's parents know closest friends' parents (1 = yes, many; 0 = other)	.41 (.49)	.42 (.49)

* T-test for differences in means significant at .05 level.

† R = respondent.

TABLE A.6.2 Means (and Standard Deviations) on Science Outcomes for African American Women by Peer Characteristics: NELS†

	Access					
Peer Variables	1988 (8th Grade) In Advanced, Enriched, or Accelerated Science Course	1990 (10th Grade) Coursework in Chemistry	1992 (12th Grade) Enrolled in Science Classes in the Last Two Years	1994 Science Major at First Postsecondary Institution	2000 First Postsecondary Degree— Science	2000 Science Degree Expected by Age 30
Importance of having strong friendships						
Very	.40 (.49)*	.19 (.39)	.86 (.35)**	.25 (.43)	.16 (.37)	.27 (.45)**
Somewhat/Not	.32 (.47)*	.18 (.39)	.93 (.25)**	.19 (.39)	.11 (.31)	.13 (.34)**
Among friends, how important to study						
Very	.44 (.50)**	.70 (.40)	.92 (.27)	.27 (.45)**	.14 (.35)	.16 (.37)**
Somewhat/Not	.31 (.46)**	.19 (.39)	.91 (.28)	.17 (.38)**	.11 (.31)	.30 (.46)**
Among friends, how important to get good grades						
Very	.44 (.50)**	.21 (.41)*	.86 (.35)**	.25 (.43)	.13 (.33)	.17 (.38)**
Somewhat/Not	.26 (.45)**	.15 (.36)*	.94 (.24)**	.19 (.39)	.13 (.33)	.36 (.48)**
Among friends, how important to finish high school						
Very	.40 (.49)**	.21 (.41)**	.91 (.28)**	.23 (.42)	.14 (.35)*	.19 (.39)**
Somewhat/Not	.23 (.42)**	.10 (.30)**	.77 (.42)**	.18 (.39)	.04 (.20)*	.50 (51)**
Among friends, how important being popular						
Very	.36 (.48)	.12 (.32)*	.74 (.44)**	.24 (.43)	.08 (.28)	.19 (.40)
Somewhat/Not	.38 (.49)	.21 (.41)*	.92 (.27)**	.22 (.42)	.14 (.34)	.25 (.44)

Among friends, how important sexual relations						
Very	.51 (.51)*	.13 (.34)	.94 (.24)	.28 (.46)	.06 (.24)	.11 (.32)
Somewhat/Not	.36 (.48)*	.20 (.40)	.88 (.33)	.21 (.41)	.13 (.36)	.26 (.44)
Importance of friends in science decision						
Very	.17 (.38)**	.11 (.33)*	‡	.46 (.52)	.15 (.37)*	.58 (.52)
Somewhat/Not	.48 (.50)**	.26 (.44)*		.37 (.49)	.24 (.43)*	.29 (.46)
Friends' desire for R after high school						
College	.37 (.48)	.23 (.42)	.89 (.32)	.24 (.43)	.15 (.36)	.16 (.37)
Not	.41 (.49)	.12 (.33)	.88 (.32)	.25 (.43)	.12 (.32)	.35 (.48)
R's parents know closest friends' parents						
Yes, Many	.36 (.48)	.22 (.42)	.82 (.38)**	.18 (.39)*	.08 (.28)*	.10 (.31)**
Other	.38 (.49)	.19 (.39)	.93 (.26)**	.27 (.44)*	.17 (.38)*	.33 (.47)**

TABLE A.6.2 *Continued*

			Achievement			
Peer Variables	1988 (8th Grade) Science Grades from 6th Grade Until Now (Mostly A's or A's & B's)	1988 (8th Grade) Science Standardized Score (Top Quartile)	1990 (10th Grade) Science Grades (Mostly A's or A's & B's)	1990 (10th Grade) Science Standardized Score (Top Quartile)	1992 (12th Grade) Science Standardized Score (Top Quartile)	2000 Current/Most Recent Occupation—Science
Importance of having strong friendships						
Very	.30 (.46)	.13 (.34)**	.36 (.48)*	.07 (.26)	.07 (.26)	.25 (.43)
Somewhat/Not	.34 (.48)	.05 (.23)**	.44 (.50)	.05 (.21)	.05 (.21)	.21 (.41)
Among friends, how important to study						
Very	.35 (.48)	.12 (.33)	.36 (.48)	.11 (.31)**	.06 (.24)	.19 (.39)**
Somewhat/Not	.31 (.46)	.11 (.31)	.40 (.49)	.04 (.20)**	.07 (.26)	.30 (.46)**
Among friends, how important to get good grades						
Very	.36 (.48)**	.10 (.30)	.27 (.45)**	.07 (.26)	.06 (.24)	.22 (.41)*
Somewhat/Not	.24 (.43)**	.12 (.33)	.54 (.50)**	.07 (.25)	.08 (.27)	.29 (.45)*
Among friends, how important to finish high school						
Very	.36 (.48)**	.13 (.34)**	.37 (.48)	.08 (.28)**	.08 (.27)*	.22 (.41)**
Somewhat/Not	.12 (.33)**	.02 (.13)**	.39 (.49)	.01 (.11)**	.02 (.16)*	.37 (.49)**
Among friends, how important being popular						
Very	.27 (.45)	.08 (.28)	.29 (.46)*	.03 (.18)*	.03 (.17)*	.10 (.31)**
Somewhat/Not	.33 (.47)	.11 (.32)	.39 (.49)*	.08 (.27)*	.07 (.26)*	.28 (.45)**

Among friends, how important sexual relations						
Very	.33 (.48)	.11 (.32)	.29 (.47)	.10 (.31)	.09 (.29)	.17 (.39)
Somewhat/Not	.32 (.47)	.11 (.31)	.38 (.49)	.07 (.25)	.06 (.24)	.25 (.43)
Importance of friends in science decision						
Very	.36 (.49)	.16 (.38)	.62 (.50)	.16 (.38)	.17 (.39)	.30 (.47)
Somewhat/Not	.33 (.47)	.24 (.43)	.49 (.50)	.16 (.37)	.13 (.34)	.29 (.46)
Friends' desire for R after high school						
College	.38 (.49)**	.14 (.35)**	.40 (.49)	.08 (.27)*	.06 (.23)	.24 (.43)
Not	.23 (.43)**	.05 (.23)**	.36 (.48)	.03 (.18)*	.05 (.23)	.28 (.45)
R's parents know closest friends' parents						
Yes, Many	.39 (.49)**	.09 (.29)	.30 (.46)**	.06 (.24)	.06 (.24)	.22 (.41)
Other	.29 (.45)**	.12 (.33)	.45 (.50)**	.08 (.27)	.07 (.26)	.27 (.45)

TABLE A.6.2 *Continued*

				Attitudes			
Peer Variables	1988 (8th Grade) Usually Look Forward to Science Class	1988 (8th Grade) Science Will Be Useful in My Future	1990 (10th Grade) Often Work Hard in Science Class	1992 (12th Grade) Interested in Science	1992 (12th Grade) Does Well in Science	1992 (12th Grade) Need Science for Job After High School	2000 Planned Occupation by Age 30— Science
Importance of having strong friendships							
Very	.59 (.49)	.68 (.47)	.69 (.47)**	.46 (.50)*	.48 (.50)	.37 (.49)**	.40 (.49)**
Somewhat/Not	.54 (.50)	.70 (.46)	.55 (.50)**	.35 (.48)*	.48 (.51)	.18 (.39)**	.17 (.38)**
Among friends, how important to study							
Very	.69 (.46)**	.71 (.46)**	.67 (.47)**	.47 (.50)	.55 (.50)	.39 (.49)**	.26 (.44)*
Somewhat/Not	.53 (.50)**	.60 (.49)**	.56 (.50)**	.36 (.49)	.39 (.49)	.16 (.38)**	.33 (.47)*
Among friends, how important to get good grades							
Very	.65 (.48)**	.72 (.45)**	.63 (.48)	.43 (.50)	.52 (.50)*	.35 (.48)*	.30 (.46)*
Somewhat/Not	.47 (.50)**	.60 (.49)**	.57 (.50)	.39 (.49)	.39 (.49)*	.19 (.40)*	.37 (.48)*
Among friends, how important to finish high school							
Very	.65 (.48)**	.69 (.46)*	.59 (.49)	.44 (.50)*	.53 (.50)**	.32 (.47)*	.25 (.43)**
Somewhat/Not	.32 (.47)**	.60 (.49)*	.68 (.47)	.25 (.45)*	.10 (.10)**	.08 (.29)	.65 (.48)**
Among friends, how important being popular							
Very	.58 (.50)	.75 (.43)*	.62 (.49)	.41 (.51)	.58 (.51)	.34 (.49)	.40 (.49)*
Somewhat/Not	.58 (.49)	.65 (.48)*	.60 (.49)	.42 (.50)	.45 (.50)	.28 (.45)	.30 (.46)*

Among friends, how important sexual relations							
Very	.66 (.48)	.64 (.49)	.56 (.51)	.29 (.48)	.35 (.50)	.27 (.48)	.28 (.46)
Somewhat/Not	.57 (.50)	.66 (.47)	.59 (.49)	.43 (.50)	.48 (.50)	.30 (.46)	.34 (.47)
Importance of friends in science decision							
Very	.79 (.46)	.79 (.42)	.70 (.42)	.28 (.46)	.57 (.51)	.31 (.48)	.40 (.50)
Somewhat/Not	.59 (.50)	.75 (.44)	.53 (.50)	.48 (.50)	.45 (.50)	.27 (.45)	.33 (.47)
Friends' desire for R after high school							
College	.61 (.49)	.72 (.45)**	.56 (.50)**	.49 (.50)*	.58 (.49)**	.38 (.49)*	.30 (.46)
Not	.55 (.50)	.60 (.49)**	.72 (.45)**	.32 (.47)*	.29 (.46)**	.18 (.40)*	.35 (.48)
R's parents know closest friends' parents							
Yes, Many	.61 (.49)	.74 (.44)**	.57 (.50)	.62 (.49)**	.68 (.47)**	.36 (.49)	.29 (.46)
Other	.55 (.50)	.62 (.49)**	.62 (.49)	.33 (.47)**	.39 (.49)**	.25 (.44)	.37 (.48)

* T-test for difference in means significant at .20 level.

** T-test for difference in means significant at .05 level.

† R = respondent.

‡ Some N's were too low for the t-test to be performed.

TABLE A.6.3 Results from Logistic Regression Models Showing Effects of Peer Variables on Science Outcomes for African American Women: NELS†

Peer Variables	Access				Achievement		Attitudes
	1992 (12th Grade) Enrolled in Science Classes in the Last Two Years	1994 Science Major at First Postsecondary Institution	2000 First Postsecondary Degree—Science	2000 Science Degree Expected by Age 30	1992 (12th Grade) Science Standardized Score (Top Quartile)	2000 Current/Most Recent Occupation—Science	2000 Planned Occupation by Age 30—Science
Importance having strong friendships (1 = very; 0 = somewhat or not)	.93 (.49)	.76 (.34)	2.34 (.51)*	1.15 (.45)	1.24 (.60)	1.07 (.30)	2.47 (.33)**
Among friends, how important to study (1 = very; 0 = somewhat or not)	.85 (.56)	1.57 (.41)	1.04 (.61)	.66 (.56)	.51 (.66)	.38 (.35)**	.94 (.37)
Among friends, how important to get good grades (1 = very; 0 = somewhat or not)	.67 (.67)	1.14 (.45)	1.11 (.67)	.54 (.59)	.98 (.69)	2.84 (.40)**	1.41 (.45)
Among friends, how important to finish high school (1 = very; 0 = somewhat or not)	.98 (.87)	.98 (.59)	2.59 (1.05)	.92 (.62)	4.44 (1.37)	.28 (.47)**	.33 (.49)**
Among friends, how important being popular (1 = very; 0 = somewhat or not)	.42 (.52)*	1.94 (.44)*	.51 (.75)	.54 (.62)	.53 (.99)	.37 (.47)**	.63 (.42)
Among friends, how important sexual relations (1 = very; 0 = somewhat or not)	15.73 (2.46)	.91 (.65)	.61 (1.22)	.75 (.97)	2.23 (.84)	.56 (.59)	.92 (.59)

Friends' desire for R after high school (1 = college; 0 = not)	1.70 (.47)	1.19 (.37)	2.20 (.64)	.71 (.42)	.95 (.59)	.99 (.30)	.75 (.31)
R's parents know closest friends' parents (1 = yes, many; 0 = other)	.52 (.47)*	.55 (.35)*	.69 (.48)	.24 (.49)**	.81 (.58)	.95 (.30)	.74 (.31)
SES	1.59 (.25)*	.91 (.16)	.83 (.23)	.75 (.21)*	2.18 (.29)**	.99 (.14)	.77 (.15)*
Constant	8.07 (.81)**	.33 (.55)**	.03 (1.13)**	1.88 (.63)	.00 (1.48)**	.96 (.45)	1.17 (.49)
Model χ^2 (df)	14.49 (9)*	9.81 (9)	7.55 (9)	34.06 (9)**	17.20 (9)**	27.10 (9)**	36.29 (9)**

* Significant at .20 level.

** Significant at .05 level.

† Table shows antilogs and standard errors in parentheses. R = respondent.

Appendix B:
Detail on Knowledge Networks
Sampling

Knowledge Networks enrolls households through a combination of mail and telephone contacts. Starting with a random sample of telephone numbers, Knowledge Networks uses multiple reverse directories to find associated addresses. If an address is available, an advance letter of introduction is sent, alerting household members to expect a telephone call within a week. Any U.S. household with a telephone has the potential to be selected for the Knowledge Networks Panel, which includes both computer users and noncomputer users. In exchange for free Internet hardware (such as a television set-top box that provides Internet access much like a desktop computer), connectivity (an Internet connection paid for by Knowledge Networks), and on-site installation, participants agree to complete a maximum of one survey per week. To trigger a survey, e-mail messages are sent to those panel members who satisfy the screening criteria for the particular study. Panel members provide all information voluntarily and with full informed consent. Each member of the participating household receives a password-protected e-mail account. Knowledge Networks maintains a call center to provide technical support and facilitate household cooperation. Selected households remain on the panel for two to three years, at which time they are eligible for retirement. At retirement, households may keep their Internet hardware equipment, but payments for Internet access are discontinued. As households retire, they are replaced with new recruits, assuring a balanced panel of consistent or growing size.

Knowledge Networks (unlike most Web-based surveys) utilizes list-assisted random digit dialing sampling techniques on a sample frame consisting of the entire United States telephone population. Sampling is done without replacement in order to ensure that numbers already fielded by Knowledge Networks do not get fielded again. Having generated the initial list of telephone numbers, the sample preparation system excludes confirmed disconnected and nonresidential telephone numbers. Next, the sample is screened to exclude numbers that are not in the WebTV Internet Service Provider network. This process results in the exclusion of only a small fraction of the U.S. population. Telephone numbers for which Knowledge Networks is able to recover a valid postal address (about 50%) are sent an advance mailing informing them that they have been selected to participate in the Knowledge Networks panel. In addition to information about the Knowledge Networks Panel, the advance mailing also contains a monetary incentive to encourage cooperation when the interviewer calls.

Following the mailing, the telephone recruitment process begins. At least fifteen dial attempts are made on unanswered calls for a period of up to ninety days. Extensive refusal conversion is also performed. All recruitment interviews are conducted by experienced interviewers. The household member is told that in return for completing a short survey weekly, the household will be given a WebTV set-top box and free monthly Internet access. All members in the household are then enumerated, and some initial demographic variables and background information of prior computer and Internet usage are collected.

Since Knowledge Networks uses telephone interviews to make initial contact with its panel households, there is potential sampling bias associated with the exclusion of nontelephone households from random digit dial (RDD) sampling frames. However, telephone coverage in African American households (95%) and white households (98%) is high and similar coverage exits in the two groups (U.S. Bureau of the Census, 2000).

In this study, respondents were randomly selected from Knowledge Network's panel of African American (and white) households with thirteen- to thirty-year-old young women. The sample includes 281 young African American women and a control group of similarly aged young white women (N = 781). These sample sizes allow us to generate reliable estimates of young African American (and white) women's science perceptions. The main survey was fielded between February 20 and May 19, 2003.

Knowledge Networks' panel acceptance rate (percent of respondents who agree to become panel members) is 37%, calculated by standards established by the American Association for Public Opinion Research (AAPOR). The within-survey completion rate—or percentage of panel members who completed the questionnaire among those who received it—was 53.7%. Research comparing data produced by Knowledge Networks' Web-based survey methodology with data produced by RDD telephone surveys suggests that Knowledge Network's parameter estimates are very similar to the estimates of RDD samples (Baker et al., 2003; Berrens, 2003; Krosnick and Chang, 2001).

The Knowledge Networks panel is representative of the U.S. population on critical demographic, geographic, economic, and social characteristics. There are four factors that help account for this. Most important, the sample is selected using list-assisted RDD telephone methods, thus creating a probability-based starting panel of U.S. telephone households. Another important factor is the panel sample weights that are adjusted to U.S. Census demographic benchmarks, reducing error associated with households without telephones and bias due to nonresponse and other nonsampling errors. Additionally, samples are selected from the panel using probability methods and appropriate design weights. Finally, nonresponse and poststratification weighting adjustments applied to the final survey data reduce effects of nonsampling error. These factors all help contribute to weighted sample estimates that differ only modestly from Census estimates on critical sociodemographic characteristics. It is also important to note that analyses of panel attrition suggest no significant differences between those who remain on the panel and those who do not (Dennis and Li, 2003). Knowledge Networks has more than 25,000 households nationwide in its current Web-enabled panel.

Appendix C:
Text for Vignettes

VIGNETTE 1: Picture of a young African American woman.

Text: "Hi, my name is LaToya. I'm an African American student and I go to high school in Northern Virginia. I love science. Especially biology. I am thinking of becoming a veterinarian. But even though I love science, I don't feel so welcome there. I mean like, the other day I was sitting in my biology class and all of a sudden I noticed that I don't get called on very often in this class. And it's like my teacher's white and most of the students in the class are white. Then I look in my textbook and see pictures of all of these white scientists. Sometimes I don't feel like I belong in science."

VIGNETTE 2: Picture of a young African American woman.

Text: "Hi, my name is LaToya. I'm an African American student and I go to high school in Northern Virginia. I love science. Especially biology. I am thinking of becoming a veterinarian. But even though I love science, I don't feel so welcome there. I mean like, the other day I was sitting in my biology class and all of a sudden I noticed that I don't get called on very often in this class. And it's like my teacher's male and most of the students in the class are male. Then I look in my textbook and see pictures of all of these male scientists. Sometimes I don't feel like I belong in science."

VIGNETTE 3: Picture of a young white woman.

Text: "Hi, my name is Michelle. I go to high school in Northern Virginia. I love science. Especially biology. I am thinking of becoming a veterinarian. But even though I love science, I don't feel so welcome there. I mean like, the other day I was sitting in my biology class and all of a sudden I noticed that I don't get called on very often in this class. And it's like my teacher's male and most of the students in the class are male. Then I look in my textbook and see pictures of all of these male scientists. Sometimes I don't feel like I belong in science."

VIGNETTE 4: Picture of a young African American woman.

Text: "Hi, my name is LaToya. I'm an African American student and I go to high school in Northern Virginia. I love science. Especially biology. I am thinking of becoming a veterinarian. But even though I love science, I don't feel so welcome there. Sometimes I don't feel like I belong in science."

Appendix D:
Selected Questions from Knowledge Networks Survey

The four vignettes used are:

GROUP 1: LaToya's (African American) picture. Mentions race as problem in science classroom.

GROUP 2: LaToya's (African American) picture. Mentions gender as problem in science classroom.

GROUP 3: Michelle's (white) picture. Mentions gender as problem in science classroom.

GROUP 4: LaToya's (African American) picture. Does not mention race or gender in complaint about science classroom.

16B. Has anything that happened to [LaToya] (if group = 1, 2, 4) [Michelle] (if group = 3) ever happened to you?

Have you ever felt this way?

Yes   No

16C. Do you think that, like [LaToya] (if group = 1, 2, 4) [Michelle] (if group = 3), many [African American] (if group = 1, 4) [female] (if group = 2, 3) students are interested in science?

Yes   No

16D. Do you think that, like [LaToya] (if group = 1, 2, 4) [Michelle] (if group = 3), many [African American] (if group = 1, 4) [female] (if group = 2, 3) students do not feel welcome in science?

Yes   No

16ddd. How about you, do you feel welcome in science?

Yes   No

16H. Do you think that your interest and abilities in science are (show [were] if s1 = 8 through 10) influenced by the feedback you get (show [got] if s1 = 8 through 10) from high school (show [middle school] if s2 = 1) teachers and students and family?[1]

Yes   No

16I. Which do you think might have a harder time in science—women or African Americans?

Women   African Americans

Appendix E:
Detail on NELS Sampling and Data

The National Educational Longitudinal Survey (NELS) data provide an excellent data source for this activity. NELS is a nationally representative, longitudinal data set collected under the auspices of the National Center for Education Statistics (2002). The base year data for NELS were collected in 1987–88 on a nationally representative sample of 24,599 8th graders (approximately aged 13), from 1,000 schools. They were interviewed again in 1990 when they were in 10th grade, in 1992 when they were high school seniors, and in 1994 when they were two years out of high school. The fourth follow-up was conducted in the year 2000 (approximately aged 25). The NELS data have been a major data source for researchers and policy-makers. The data include extensive information on a wide variety of science education experiences. Analyses are based on those who participated in all three survey years. Those who dropped out of high school are not included. All analyses include weights which control for sample attrition and non-response. Since NELS weights were created to project to the total U.S. high school youth population, we adjusted the weights back to sample size. The sample used here consists of 581 young African American women and 3,365 young white women.

The NELS data is a valuable secondary data tool for showing the science experiences of representative samples of U.S. youth. The NELS data are frequently used to guide educators and policymakers. With the most recent (2000) panel of information on the NELS cohort, the

NELS provides the richest source of information available on the science experiences of young people from age 13 (middle school) through the high school, undergraduate, and postcollege years.

NELS has excellent longitudinal data (over a twelve-year period) on the science experiences of a cohort of young people who were in 8th grade in 1988. In this research, we do not include the social sciences in our definition of science. Majors and occupations involving physical science, mathematics, medical science, computer science, engineering, and technology are included. Three of the most important aspects of science experience include science access, science achievement, and science attitudes (Hanson 1996). NELS includes extensive measures of each. For example, measures of access include enrollment in accelerated science (8th grade), chemistry (10th grade), university/college physics classes (two years out of high school), and science major or degree (including those in math, engineering, or technology) since leaving high school (two years out of high school and eight years out of high school). Measures of achievement include (among others) grades in science and standardized science test scores during the high school years as well as science occupations in the post–high-school years. Measures of attitudes include (among others) responses to questions about looking forward to science classes, feeling challenged in these classes, being interested and doing well in science as well as others which ask about the importance of science in the respondent's future (measured during the high school years). A question on the occupation that the respondent plans to have at age 30 (coded here as science or not science) is also included in the attitudinal questions.

 Notes

Chapter 1: Introduction

1. Social sciences are not included in the definition of science used here. In a number of sections where we provide information from our analyses of national data, we refer to science majors (postsecondary) and occupations. In general, this refers to a rather broad definition of science, including physical science, mathematics, medical science, computer science, engineering, and technology. Sometimes this area of study and work is referred to as SMET (science, math, engineering, and technology) since these are interrelated areas of study and work.

2. The term "gender" will often be used in this volume. The term suggests social learning and structures that vary for men and women. For ease of distinction, I sometimes use the more demographic term "sex," but I am referring to socially created (not biological) differences between men and women in this research on science experiences.

Chapter 2: The Conceptual Framework

1. Agency is a term used by critical theorists to denote the degree of free will exerted by the individual in social actions. Structure, on the other hand, refers to recurring patterns of behavior in an area of life that puts pressure on us to conform (Jenks, 1998).

2. The African American family system has traditionally been one based on a strong extended and fictive kinship network (Roschelle, 1999).

Chapter 3: Young African American Women's Experiences in Science

1. See Appendix E for a more detailed discussion of the NELS data and sampling design.

2. Tables are provided in Appendix A.

3. We should note that the NELS samples used here and in the Hanson and Palmer-Johnson (2000) paper are slightly different. Thus, specific numbers are not directly comparable. However, the overall trends from the two studies can be contrasted with this note of caution.

4. See Appendices B, C, and D for more information on Knowledge Networks sampling, vignettes, and survey questions.

Chapter 4: Influences: Teachers and Schools

1. For the sake of parsimony, a subset of variables from each of the categories (school, family, peer) were included in these analyses.

Appendix D: Selected Questions from Knowledge Networks Survey

1. Variable s1 measures grade in school. Those who attended high school in the past but are not currently enrolled are in categories 8 through 10. Those who are currently in high school are in categories 2 through 7. Those who are in middle school are in category 1.

References

Agger, B. 1998. *Critical Social Theories: An Introduction*. Boulder, Colo.: Westview Press.

Ainsworth-Darnell, J. W., and D. B. Downey. 1998. Assessing the oppositional culture explanation for racial/ethnic differences in school performance. *American Sociological Review* 63 (4): 536–553.

Alexander, C. N., Jr., and E. W. Campbell. 1964. Peer influences on adolescent educational aspirations and attainments. *American Sociological Review* 29 (4): 568–575.

Alexander, C. S., and H. J. Becker. 1978. The use of vignettes in survey research. *Public Opinion Quarterly* 42 (1): 93–104.

Allen, P. 1997. Black students in "ivory towers." *Studies in the Education of Adults* 29 (2): 179–190.

Allison, P. D. 1984. *Event History Analysis: Regression for Longitudinal Data*. Newbury Park, Calif.: Sage.

American Association of University Women (AAUW). 1992. How schools short change girls. Washington, D.C.: American Association of University Women Educational Foundation.

Andersen, M. L. 1997. *Thinking About Women: Sociological Perspectives on Sex and Gender*. Boston: Allyn and Bacon.

Andersen, M. L., and P. H. Collins. 1995. *Race, Class, and Gender: An Anthology*. Belmont, Calif.: Wadsworth.

Anderson, T. 1988. Black encounter of racism and elitism in white academe: A critique of the system. *Journal of Black Studies* 18 (3): 259–272.

Aquilno, W. S., and L. A. Lo Sciuto. 1990. Effects of interview mode on self-reported drug use. *Public Opinion Quarterly* 54 (3): 362–395.

Atwater, M. M., J. Wiggins, and C. M. Garner. 1995. A study of urban middle school students with high and low attitudes toward science. *Journal of Research in Science Teaching* 32 (6): 665–677.

Axelson, L. J. 1970. The working life: Differences in perception among Negro and white males. *Journal of Marriage and the Family* 32 (2): 197–214.

Baker, D. R. 1987. The influence of role-specific self-concept and sex-role identity on career choices in science. *Journal of Research in Science Teaching* 24 (8): 739–756.

Baker, D., and R. Leary. 1995. Letting girls speak out about science. *Journal of Research in Science Teaching* 32 (1): 3–27.

Baker, L. M., K. Bundorf, S. Singer, and T. Wagner. 2003. *Validity of Survey of Health and Internet and Knowledge Network's Panel and Sampling.* Stanford, Calif.: Stanford University Press.

Banks, J. A. 1991. Social studies, ethnic diversity, and social change. In C. V. Willie, A. M. Garibaldi, and W. L. Reed, eds., *The Education of African-Americans,* 129–147. New York: Auburn House.

Beal, C. R. 1994. *Boys and Girls: The Development of Gender Roles.* New York: McGraw-Hill.

Bechtel, H. K. 1989. Introduction. In W. J. Pearson Jr. and H. K. Bechtel, eds. *Blacks, Science, and American Education,* 1–20. New Brunswick, N.J.: Rutgers University Press.

Bennett, P. R., and Y. Xie. 2003. Revisiting racial differences in college attendance: The role of historically black colleges and universities. *American Sociological Review* 68 (4): 567–580.

Berndt, T. J., A. E. Laychak, and K. Park. 1990. Friends' influence on adolescents' academic achievement motivation: An experimental study. *Journal of Educational Psychology* 82 (4): 664–670.

Berrens. R. 2003. The advent of internet surveys for political research: A comparison of telephone and internet samples. *Political Analysis* 11 (1): 1–22.

Berryman, S. E. 1983. *Who Will Do Science?* New York: Rockefeller Foundation.

Betz, N. 1997. What stops women and minorities from choosing and completing majors in science and engineering? In D. Johnson, ed., *Minorities and Girls in School: Effects on Achievement and Performance,* 105–140. Thousand Oaks, Calif.: Sage.

Billingsley, A. 1968. *Black Families in White America.* Englewood Cliffs, N.J.: Prentice-Hall.

Borman, G., and L. Overman. 2004. Academic resilience in mathematics among poor and minority students. *The Elementary School Journal* 104 (3): 177–195.

Boswell, S. L. 1985. The influence of sex-role stereotyping on women's attitudes and achievement in mathematics. In S. F. Chipman, L. R. Brush, and D. M. Wilson, eds., *Women and Mathematics: Balancing the Equation,* 175–197. Hillsdale, N.J.: Lawrence Erlbaum Associates.

Bourdieu, P. 1973. Cultural reproduction and social reproduction. In E. Brown, ed., *Knowledge, Education, and Cultural Change*, 71–112. London: Tavistock.

Bowman Damico, S., and C. Sparks. 1986. Cross-group contact opportunities: Impact on interpersonal relationships in desegregated middle schools. *Sociology of Education* 59 (2): 113–123.

Brown, A. H. 2000. Creative pedagogy to enhance the academic achievement of minority students in math. In S. T. Gregory, ed., *The Academic Achievement of Minority Students*, 365–390. New York: University Press of America.

Brown, B. 1990. Peer groups and peer cultures. In S. Feldman and G. Elliot, eds., *At the Threshold: The Developing Adolescent*, 171–196. Cambridge, Mass.: Harvard University Press.

Brown, R. W. 1990. Research apprenticeships for young undergraduates. In S. Z. Keither and P. Keith, eds., *Proceedings of the National Conference on Women in Mathematics and the Sciences*. St. Cloud, Minn.: St. Cloud State University Press.

Buckley, T. R., and R. T. Carter. 2005. Black adolescent girls: Do gender role and racial identity impact their self-esteem? *Sex Roles* 53 (9–10): 647–661.

Burbridge, L. B. 1991. The interaction of race, gender, and socioeconomic status in education outcomes. Working paper No. 246. Wellesley, Mass.: Center for Research on Women, Wellesley College.

Burger, C. J., and M. L. Sandy. 1998. A guide to gender fair education in science and mathematics. Office of Educational Research and Improvement, R168R5009-97. Washington, D.C.: U.S. Department of Education.

Campbell, E. Q., and C. N. Alexander. 1965. Structural effects and interpersonal relationships. *American Sociological Review* 71 (3): 284–289.

Carwell, H. 1977. *Blacks in Science: Astrophysicist to Zoologist*. Hicksville, N.Y.: Exposition Press.

Catalyst, Inc. 1992. *Women in Engineering: An Untapped Resource*. New York: Catalyst, Inc.

Catsambis, S. 1995. Gender, race, ethnicity, and science education in the middle grades. *Journal of Research in Science Teaching* 32 (3): 243–257.

———. 1994. The path to math: Gender and racial-ethnic differences in mathematics participation from middle school to high school. *Sociology of Education* 67 (3): 199–215.

Chafetz, J. S. 1984. *Sex and Advantage: A Comparative, Macro-structural Theory of Sex Stratification*. Totowa, N.J.: Rowman & Allanheld.

———. 1999. The varieties of gender theory in sociology. In J. S. Chafetz, ed., *Handbook of the Sociology of Gender*, 3–24. New York: Springer.

Cho, D. 2007. The role of high school performance in explaining women's rising college enrollment. *Economics of Education Review* 26 (4): 450–462.

Cicade, M. 2004. Volunteerism and science among young African American women. Working Paper. Department of Sociology, Catholic University of America.

Clark, R., R. R. Dogan Jr., and N. J. Akbar. 2003. Youth and parental correlates of externalizing symptoms, adaptive functioning, and academic performance: An exploratory study in preadolescent blacks. *Journal of Black Psychology* 29 (2): 210–229.

Clewell, B. C., and B. Anderson. 1991. *Women of Color in Mathematics, Science, and Engineering: A Review of the Literature.* Washington, D.C.: Center for Women Policy Studies.

Cobb, J. P. 1993. A life in science: Research and service. In C. Fort, J. Bird, and J. Didion, eds., *A Hand Up: Women Mentoring Women in Science.* Washington, D.C.: The Association for Women in Science.

Cole, J. R. 1987. *Fair Science: Women in the Scientific Community.* New York: Columbia University Press.

Cole, J. R., and S. Cole. 1973. *Social Stratification in Science.* Chicago: University of Chicago Press.

Coleman, J., E. Q. Campbell, C. J. Hobson, J. McPartland, A. M. Mood, F. D. Weinfeld, and R. L. York. 1966. *Equality of Educational Opportunity.* Washington, D.C.: U.S. Department of Health, Education and Welfare, Office of Education.

Collins, P. H. 1990a. *Black Feminist Thought.* Boston: Unwin Hyman.

———. 1990b. The meaning of motherhood in black culture. In P. H. Collins, ed., *Black Feminist Thought,* 119–132. Boston: Unwin Hyman.

———. 1999. Moving beyond gender: Intersectionality and scientific knowledge. In M. M. Ferree, J. Lorber, and B. B. Hess, eds., *Revisioning Gender,* 261–284. Thousand Oaks, Calif.: Sage.

Connell, R. W. 1987. *Gender and Power: Society, the Person, and Politics.* Stanford, Calif.: Stanford University Press.

Constant, E. O. 1989. Science in society: Petroleum engineers and the oil fraternity in Texas. *Social Studies of Science* 9 (3): 439–472.

Cooper, R. 1996. Detracking reform in an urban California high school: Improving the schooling experiences of African American students. *The Journal of Negro Education* 65 (2): 190–208.

Corsaro, W. A., and D. Eder. 1990. Children's peer cultures. *Annual Review of Sociology* 16: 197–220.

Creswell, J. L., and R. H. Exezidis. 1982. Research brief: Sex and ethnic differences in mathematics achievement of Black and Mexican American adolescents. *Texas Tech Journal of Education* 9 (3): 219–222.

Cross, T., and R. B. Slater. 2001. The troublesome decline in African American college student graduation rates. *Journal of Blacks in Higher Education* 33 (Autumn): 102–109.

Cummins, J. 1993. Empowering minority students: A framework for intervention. In L. Weiss and M. Fine, eds., *Beyond Silenced Voices: Class, Race, and*

Gender in United States Schools, 101–117. Albany: State University of New York Press.

Davis, C. G., B. F. Sloat, N. G. Thomas, and J. D. Manis. 1989. An analysis of factors affecting choice of majors in science, mathematics, and engineering at the University of Michigan. Ann Arbor, Mich.: University of Michigan Center for Education of Women Research Reports.

Davis, K. S. 1999. Why science? Women scientists and their pathways along the road less traveled. *Journal of Women and Minorities in Science and Engineering* 5 (2): 129–153.

Davis, R. D. 2004. *Black Students' Perceptions: The Complexity of Persistence to Graduation at an American University.* New York: Peter Lang.

Dennis, J. M., and R. Li. 2003. Effects of panel attrition on survey estimates. Paper presented at the annual meeting of the American Association for Public Opinion Research. Nashville, Tenn. (May).

Domhoff, G. W. 1983. *Who Rules America Now?* Englewood Cliffs, N.J.: Prentice-Hall.

Downey, D. B., and S. Pribesh. 2004. When race matters: Teachers' evaluations of students' classroom behavior. *Sociology of Education* 77 (4): 267–282.

Dugger, K. 1988. Social location and gender-role attitudes: A comparison of black and white women. *Gender and Society* 2 (4): 425–448.

Duran, R. P. 1987. Hispanics' pre-college and undergraduate education: Implications for science and engineering studies. In L. S. Dix, ed., *Minorities: Their Under-representation and Career Differentials in Science and Engineering,* 73–128. Washington, D.C.: National Academy Press.

Eccles, J. 1997. User friendly science and mathematics: Can it interest girls and minorities in breaking through the middle school wall? In D. Johnson, ed., *Minorities and Girls in School: Effects on Achievement and Performance,* 65–104. Thousand Oaks, Calif.: Sage.

Economic and Social Research Council. 2008. Math plus "geeky" image equals deterred students. *Science Daily.* http://www.sciencedaily.com/releases/2008/05/080512094435.htm (accessed July 8, 2008).

Edwards, L. D. 1999. Using multimedia to counter stereotypes in science classrooms: New perceptions on who becomes a scientist. Paper presented at the annual convention of the Association for Educational Communications and Technology. Houston, Texas (February).

Eitzen, D. S., and M. B. Zinn. 2004. *In Conflict and Order.* Boston: Allyn & Bacon.

Erb, T. O., and W. Smith. 1984. Validation of the attitude toward women in science scale for early adolescents. *Journal of Research in Science Teaching* 21 (4): 391–397.

Eshelman, J. R. 2002. *The Family.* New York: Allyn and Bacon.

Evans, G. 1988. Those loud black girls. In D. Spender and E. Sarah, eds., *Learning to Lose: Sexism and Education,* 183–190. London: The Women's Press.

Farenger, S. J., and B. A. Joyce. 1999. Intentions of young students to enroll in science courses in the future: An examination of gender differences. *Science and Education* 83 (1): 55–75.

Feagin, J. R., H. Vera, and N. Imani. 1996. *The Agony of Education: Black Students at White Colleges and Universities.* New York: Routledge.

Felice, L. G. 1981. Black student dropout behavior: Disengagement from school, rejection, and racial discrimination. *Journal of Negro Education* 50 (4): 415–424.

Feree, M. M. 1990. Beyond separate spheres: Feminism and family research. *Journal of Marriage and the Family* 52 (4): 866–884.

Fish, V. K. 1979. Where are the women scientists: The role of parents, teachers, and friends in the self-concept process. ERIC document number ED193064. Washington, D.C.: U.S. Department of Education.

Fordham, S. 1993. "Those loud black girls": (Black) women, silence, and gender "passing" in the academy. *Anthropology and Education Quarterly* 24 (1): 3–32.

Fordham, S., and J. U. Ogbu. 1986. Black students' school success: Coping with the burden of "acting white." *Urban Review* 18 (3): 176–206.

Foster, M. 1993. Resisting racism: Personal testimonies of African American teachers. In L. Weiss and M. Fine, eds., *Beyond Silenced Voices: Class, Race, and Gender in United States Schools,* 273–288. Albany: State University of New York Press.

Foster, M., and L. R. Perry. 1982. Self-evaluation among blacks. *Social Work* 27 (1): 60–66.

Fox, L. H., L. Brody, and D. Tobin. 1985. The impact of early intervention programs upon course-taking and attitudes in high school. In S. F. Chipman, F. R. Brush, and D. M. Wilson, eds., *Women and Mathematics: Balancing the Equation,* 249–274. Hillsdale, N.J.: Lawrence Erlbaum Associates.

Fox, L. H., and D. Tobin, eds. 1980. *Women and the Mathematical Mystique.* Baltimore: Johns Hopkins University Press.

Fox, L. H., D. Tobin, and L. Brody. 1979. Sex role socialization and achievement in mathematics. In M. A. Wittig and A. C. Petersen, eds., *Sex-Related Differences in Cognitive Functioning,* 303–332. New York: Academic Press.

Fredricks, J. A., and J. S. Eccles. 2006. Developmental benefits of extracurricual involvement: Do peer characteristics mediate the link between activities and youth outcomes? *Journal of Youth Adolescence* 34 (6): 507–520.

Freeman, K. 1997. Increasing African American's participation in higher education: African American high school students' perspectives. *The Journal of Higher Education* 68 (5): 523–550.

Fries-Britt, S. 1998. Moving beyond black achiever isolation: Experiences of gifted black collegians. *The Journal of Higher Education* 69 (5): 556–576.

Fuchs Epstein, C. 1973. Positive effects of the multiple negative: Explaining the success of black professional women. *American Journal of Sociology* 78 (4): 912–935.

Furman, W., and D. Buhrmester. 1992. Age and sex differences in perceptions of networks of personal relationships. *Child Development* 63 (1): 103–115.

Gaskell, J. 1985. Course enrollment in the high school: The perspective of working class females. *Sociology of Education* 58 (1): 48–59.

Ginwright, S. A. 2004 *Black in School: Afrocentric Reform, Urban Youth, and the Promise of Hip-Hop Culture*. New York: Teachers College, Columbia University.

Giroux, H. A. 1983. *Theory and Resistance in Education: A Pedagogy for the Opposition*. Hadley, Mass.: Bergin and Garvey.

Glenn, E. N. 1985. Racial ethnic women's labor: The intersection of race, gender, and class oppression. *Review of Radical Political Economics* 17 (3): 8–108.

Goodenow, C., and K. E. Grady. 1993. The relationship of school belonging and friends' values to academic motivation among urban adolescent students. *Journal of Experimental Education* 62 (1): 60–71.

Grandy, J. 1998. Persistence in science of high-ability minority students: Results of a longitudinal study. *The Journal of Higher Education* 69 (6): 589–620.

Grant, L., P. M. Horan, and B. Watts-Warren. 1994. Theoretical diversity in the analysis of gender and education. *Research in Sociology of Education and Socialization* 10: 71–109.

Greenfield, T. A. 1996. Gender, ethnicity, science achievement, and attitudes. *Journal of Research on Science Teaching* 33 (8): 901–933.

Gump, J. 1975. Comparative analysis of black women's and white women's sex-role attitudes. *Journal of Consulting and Clinical Psychology* 43 (6): 858–863.

Gustafson, S., H. Stattin, and D. Magnusson. 1992. Aspects of the development of a career versus homemaking orientation among females: The longitudinal influence of educational motivation and peers. *Journal of Research on Adolescence* 2 (3): 241–259.

Gutman, H. G. 1976. *The Black Family in Slavery and Freedom*. New York: Vintage.

Haller, A. O., and C. E. Butterworth. 1960. Peer influences on levels of occupational and educational aspiration. *Social Forces* 38 (4): 289–295.

Haney, W. 1993. Testing and minorities. In L. Weiss and M. Fine, ed., *Beyond Silenced Voices: Class, Race, and Gender in United States Schools*, 45–73. Albany: State University of New York Press.

Hanson, S. L. 1994. Lost talent: Unrealized educational aspirations and expectations among U.S. youth. *Sociology of Education* 67 (3): 159–183.

———. 1996. *Lost Talent: Women in the Sciences*. Philadelphia: Temple University Press.

————. 2004. African American women in science: Experiences from high school through the post-secondary years and beyond. *National Women's Studies Association Journal* 16 (1): 96–115.

————. 2006a. Success in science among young African American women: The role of minority families. *Journal of Family Issues* 28 (1): 3–33.

————. 2006b. Insights from vignettes: African American women's perceptions of discrimination in the science classroom. *Journal of Women and Minorities in Science and Engineering* 12 (1): 11–34.

Hanson, S. L., S. Fuchs, S. Aisenbrey, and N. Kravets. 2004. Attitudes toward gender, work, and family among male and female scientists in Germany and the United States. *Journal of Women and Minorities in Science and Engineering* 10 (2): 99–129.

Hanson, S. L., and A. L. Ginsburg. 1988. Gaining ground: Values and high school success. *American Educational Research Journal* 25 (3): 334–365.

Hanson, S. L., and R. S. Kraus. 1998. Women, sport, and science: Do female athletes have an advantage? *Sociology of Education* 71 (2): 93–110.

————. 1999. Women in male domains: Sport and science. *Sociology of Sport Journal* 16 (2): 92–110.

Hanson, S. L., and E. Palmer-Johnson. 2000. Expecting the unexpected: A comparative study of African American women's experiences in science during the high school years. *Journal of Women and Minorities in Science and Engineering* 6 (4): 265–294.

Harding, S. 1986. *The Science Question in Feminism*. Ithaca, N.Y.: Cornell University Press.

Harlen, W. 1985. Girls and primary-school science education: Sexism, stereotypes and remedies. *Prospects: Quarterly Review of Education* 15 (4): 541–551.

Hart, L. E., and G. M. A. Stanic. 1989. Attitudes and achievement-related behaviors of middle school mathematics students: Views through four lenses. Paper presented at the annual meetings of the American Educational Research Association. San Francisco (March).

Higginbotham, E., and L. Weber. 1992. Moving up with kin and community: Upward social mobility for black and white women. *Gender and Society* 6 (3): 416–40.

Hill, R. B. 1971. *The Strengths of Black Families*. New York: Emerson Hall.

Hill, S. A., and J. Sprague. 1999. Parenting in black and white families: The interaction of gender and class and race. *Gender and Society* 13 (4): 480–502.

Holland, D. C., and M. S. Eisenhart. 1991. *Educated in Romance: Women, Achievement, and College Culture*. Chicago: University of Chicago Press.

Hoyte, R. M., and J. Collett. 1993. "I can do it": Minority undergraduate science experiences and the professional career choice. In J. Gainen and R. Boice, eds., *Building a Diverse Faculty*, 81–91. San Francisco: Jossey-Bass.

Hueftle, J. J., S. J. Rakow, and W. W. Welch. 1983. *Images of Science*. Minneapolis: University of Minnesota, Science Assessment and Research Project.

Ide, J. K., J. Parkerson, G. D. Haertel, and H. Walbert. 1981. Peer group influence on educational outcomes: A quantitative synthesis. *Journal of Educational Psychology* 73 (4): 472–484.

Jackson, J. 2005. *Race, Racism, and Science: Social Impact and Interaction.* New Brunswick, N.J.: Rutgers University Press.

Jacobowitz, T. J. 1983. Relationship of sex, achievement, and science self-concept to the science career preferences of black students. *Journal of Research in Science Training* 20 (7): 621–628.

Jacobs, J. E., L. L. Finken, N. L. Griffin, and J. D. Wright. 1998. The career plans of science-talented rural adolescent girls. *American Educational Research Journal* 35 (4): 681–704.

Jeffe, D. B. 1995. About girls' "difficulties" in science: A social, not a personal matter. *Teachers College Record* 97 (2): 206–226.

Jenks, C., ed. 1998. *Core Sociological Dichotomies.* Thousand Oaks, Calif.: Sage.

Johnson, S. D. 1997. The multiple hats of identity: Addressing the various components of African American identity in student development. *College Student Affairs Journal* 16 (2): 65–72.

Jordan, D. 1999. Black women in the agronomic sciences: Factors influencing career development. *Women and Minorities in Science and Engineering* 5 (2): 113–128.

Journal of Blacks in Higher Education. 2001. It's the strong academic performance of African-American women that accounts for the closing of the income gap between college-educated blacks and whites. Commentary. http://www.jstor.org/jstor/gifcvtdir/di001737/10773771/di020320/02p0432z_1.1gif?jstor (accessed June 11, 2007).

Jovanovic, J. U., and S. S. King. 1998. Boys and girls in the performance-based science classroom: Who's doing the performing? *American Educational Research Journal* 35 (3): 477–496.

Kahle, J. B., and M. Lakes. 1983. The myth of equality in science classrooms. *Journal of Research on Science Teaching* 20 (2): 131–140.

Kane, E. W. 2000. Racial and ethnic variations in gender-related attitudes. *Annual Review of Sociology* 26: 419–439.

Kantner, R. M. 1977. *Men and Women of the Corporation.* New York: Basic.

Kao, G., and J. S. Thompson. 2003. Racial and ethnic stratification in educational achievement and attainment. *Annual Review of Sociology* 29: 417–442.

Kelly, A. 1988. Option choice for girls and boys. *Research in Science and Technological Education* 6 (1): 5–23.

Kenschaft, P. C. 1991. *Winning Women into Mathematics.* Washington, D.C.: Mathematical Association of America.

———. 1981. Black women in mathematics in the United States. *American Mathematical Monthly* (October): 592–604.

Kerr, B. 2000. Guiding math/science talented girls and women. NSF Award HRD, EHR 9619121. Arlington, Va.: National Science Foundation.

Kinney, D. A. 1993. From nerds to normals: The recovery of identity among adolescents from middle school to high school. *Sociology of Education* 66 (1): 21–40.

Krosnick, J., and L. C. Chang. 2001. A comparison of random digit dialing telephone survey telephone methodology with internet survey methodology as implemented by Knowledge Networks and Harris Interactive. Paper presented at the Annual Conference of the American Association for Public Opinion Research. Montreal (May).

LaFollette, M. C. 1988. Eyes on the stars: Images of women scientists in popular magazines. *Science, Technology and Human Values* 13 (3–4): 262–275.

Leaman, O., and B. Carrington. 1985. Athleticism and the reproduction of gender and ethnic marginality. *Leisure Studies* 4 (2): 205–217.

Lee, J. D., C. Stow, and S. Nelson. 2003. If I dropped out, they'd be shocked: Exploring factors that promote retention or loss of talented science and technology college students. Paper presented at the annual meetings of the Southern Sociological Society. Atlanta, Ga. (April).

Leggon, C. B. 2003. Women of color in IT: Degree trends and policy implications. *IEEE Technology and Society Magazine* (Fall): 36–42.

Leggon, C. B., and W. Pearson Jr. 1997. The baccalaureate origins of African American female Ph.D. scientists. *Journal of Women and Minorities in Science and Engineering* 3 (4): 213–224.

Levitt, J. J., J. L. Levitt, G. L. Bustos, N. A. Crooks, J. D. Santos, P. Telan, and M. E. Silver. 1999. The social ecology of achievement in pre-adolescents: Social support and school attitudes. Paper presented at the annual meeting of the American Educational Research Association. Montreal (April).

Liontos, L. B. 1991. Involving at-risk families in their children's education (EDD00036). Washington, D.C.: U.S. Department of Education, Office of Educational Research and Improvement.

Lomax, R. G., M. M. West, M. C. Harmon, K. A. Viator, and G. F. Madaus. 1995. The impact of mandated standardized testing on minority students. *The Journal of Negro Education* 64 (2): 171–185.

Lopez, N. 2003. *Hopeful Girls, Troubled Boys: Race and Gender Disparity in Urban Education*. New York: Routledge.

Lorber, J. 1994. *Paradoxes of Gender*. New Haven, Conn.: Yale University Press.
———. 2001. *Gender Inequality: Feminist Theories and Politics*. Los Angeles: Roxbury.

Luster, T., and H. P. McAdoo. 1995. Factors related to self-esteem among African American youths: A secondary analysis of the High/Scope Perry Preschool data. *Journal of Research on Adolescence* 5 (4): 451–467.

Lyall, S. 1987. A women's college looks to men, skeptically, for survival. *New York Times*, April 26, E8.

Maccorquodale, P. 1984. Self-image, science, and math: Does the image of the "scientist" keep girls and minorities from pursuing science and math? Paper presented at the annual meeting of the American Sociological Association. San Antonio, Texas (August).

Maholmes, V. 2001. Revisiting stereotype threat: Examining minority students' attitudes toward learning mathematics and science. *Race, Gender, & Class* 8 (1): 8–21.

Malcom, S. M. 1976. *The Double Bind: The Price of Being a Minority Woman in Science.* Washington, D.C.: American Association for Advancement of Science.

———. 1983. *An Assessment of Programs that Facilitate Increased Access and Achievement of Females and Minorities in K–12 Mathematics and Science Education.* Washington, D.C.: American Association for the Advancement of Science.

Malcom, S. M., V. V. Van Horne, and C. Gaddy. 1998. Losing ground: Science and engineering graduate education of black Americans. Report for the American Association for the Advancement of Science.

Manis, J. D., N. G. Thomas, B. F. Sloat, and C. G. Davis. 1989. An analysis of factors affecting choice of majors in science, mathematics, and engineering at the University of Michigan. Ann Arbor: University of Michigan Center for Education of Women Research Reports.

Maple, S. A., and F. K. Stage. 1991. Influences on the choice of math/science major by gender and ethnicity. *American Educational Research Journal* 28 (1): 37–60.

Margolis, J., J. J. Holme, R. Estrella, J. Goode, K. Nao, and S. Stumme. 2003. The computer science pipeline in urban high schools: Access to what? For whom? *IEEE Technology and Society Magazine* (Fall): 12–48.

Marrett, C. B. 1981. Patterns of enrollment in high school mathematics and science: Final report. Madison: Wisconsin Center for Education Research.

———. 1982. Minority females in high school mathematics and science. Report from the Program in Student Diversity and School Processes, Wisconsin Center for Education Research, Madison. Washington, D.C.: National Institute of Education.

Matthews, S. W. 1980. Race and sex-related difference in high school mathematics enrollment. Ph.D. dissertation, University of Chicago. *Dissertation Abstracts International,* 41 (9): 3934-A.

———. 1984. Influences on the learning and participation of minorities in mathematics. *Journal for Research in Mathematics Education* 15 (2): 84–95.

Matyas, M. L. 1986. Persistence in science-oriented majors: Factors related to attrition among male and female students. Paper presented at the annual meetings of the American Educational Research Association. San Francisco, Calif. (March).

Mau, W., M. Domnick, and R. A. Ellsworth. 1995. Characteristics of female students who aspire to science and engineering or homemaking occupations. *The Career Development Quarterly* 43 (4): 323–337.

McCubbin, H. I., E. A. Thompson, and J. A. Futrell, eds. 1998. *Resiliency in African-American Families.* Thousand Oaks, Calif.: Sage.

McDuffie, T. E., Jr. 2001. Scientists—Geeks and nerds? *Science and Children* 38 (May): 16–19.

McLean, J. A., et al. 2000. EPWG: The Turnage science, math and technology scholars program. NSF Award HRD, EHR 9555817. Arlington, Va.: National Science Foundation.

McNamara Horvat, E., and K. S. Lewis. 2003. Reassessing the "burden of acting white": The importance of peer groups in managing academic success. *Sociology of Education* 76 (4): 265–280.

McNeal, R. B., Jr. 1999. Participation in high school extracurricular activities: Investigating school effects. *Social Science Quarterly* 80 (2): 291–309.

Mickelson, R. A. 1990. The attitude-achievement paradox among black adolescents. *Sociology of Education* 63 (1): 44–61.

Mills, C. W. 1956. *The Power Elite*. New York: Oxford University Press.

MIT (Massachusetts Institute of Technology). 1999. *A Study on the Status of Women Faculty in Science at MIT*. The MIT Faculty Newsletter (XI:4), March. Boston: MIT.

Moses, Y. T. 1989. *Black Women in Academe: Issues and Strategies*. Washington, D.C.: Association of American Colleges and Universities.

Moynihan, D. P. 1965. *The Negro Family: The Case for National Action*. Washington, D.C.: U.S. Department of Labor.

Murray, S. R., and M. Mednick. 1977. Black women's achievement orientation: Motivation and cognitive factors. *Psychology of Women Quarterly* 1 (3): 247–249.

Murrel, A. J., I. H. Frieze, and J. L. Frost. 1991. Aspiring to careers in male- and female-dominated professions. *Psychology of Women Quarterly* 15 (1): 103–126.

Musick, M., W. Bynum, and J. Wilson. 2000. Race and formal volunteering: The differential effects of class and religion. *Social Forces* 78 (4): 1539–1571.

National Center for Education Statistics. 2000a. *Entry and Persistence of Women and Minorities in College Science and Engineering Education*. NCES 2000-601, by G. Huang, N. Taddese, and E. Walter. Washington, D.C.: U.S. Department of Education.

———. 2000b. *Trends in Educational Equity of Girls & Women*. NCES 2000-030, by Y. Bae, S. Choy, C. Geddes, J. Sable, and T. Snyder. Washington, D.C.: U.S. Department of Education.

National Science Board. 2000. *Science and Engineering Indicators—2000*. Arlington, Va.: National Science Foundation.

National Science Foundation. 1999. *Women, Minorities, and Persons with Disabilities in Science and Engineering, 1998*. Arlington, Va.: National Science Foundation.

———. 2000. *Women, Minorities, and Persons with Disabilities in Science and Engineering, 2000*. Arlington, Va.: National Science Foundation.

———. 2004. *Women, Minorities, and Persons with Disabilities in Science and Engineering, 2004*. Arlington, Va.: National Science Foundation.

Oakes, J. 1990. *Lost Talent: The Underparticipation of Women, Minorities, and Disabled Persons in Science*. Santa Monica: Rand.

Oakes, J., and M. Lipton. 1992. Detracking schools: Early lessons from the field. *Phi Delta Kappa* 73 (6): 448–454.

Ogbu, J. U. 1974. *The Next Generation: An Ethnography of Education in an Urban Neighborhood.* New York: Academic Press.

———. 1978. *Minority Education and Caste.* New York: Academic Press.

———. 1985. Research currents: Cultural-ecological influences on minority school learning. *Language Arts* 62 (December): 860–869

———. 1986. The consequences of the American caste system. In U. Neisser, ed., *The School Achievement of Minority Children,* 19–56. Hillsdale, N.J.: Lawrence Erlbaum Associates.

———. 1991. Low performance as an adaptation: The case of blacks in Stockton, California. In M. A. Gibson and J. U. Obgu, eds., *Minority Status and Schooling,* 249–285. New York: Garland.

Olsen, C. D. 1996. African American adolescent women: Perceptions of gender, race, and class. *Marriage and the Family Review* 21 (1–2): 105–121.

Osmond, M. W., and B. Thorne. 1993. Feminist theories: The social construction of gender in families and society. In P. G. Boss, W. J. Doherty, R. LaRossa, W. R. Schumm, and S. K. Steinmetz, eds., *Sourcebook of Family Theory and Methods: A Contextual Approach,* 591–623. New York: Plenum Press.

Ovelton Sammons, V. 1990. *Blacks in Science and Medicine.* Washington, D.C.: Hemisphere Publishing Corp.

Palmer, D. H. 1997. Investigating students' private perceptions of scientists and their work. *Research in Science and Technological Education* 15 (2): 173–183.

Pearson, W., Jr. 1978. Race and universalism in the scientific community. *Sociological Inquiry* 48 (1): 38–53.

———. 1982. The social origins of black American doctorates in the sciences. *Sociological Spectrum* 2 (1): 13–29.

———. 1985. *Black Scientists, White Society, and Colorless Science: A Study of Universalism in American Science.* New York: Associated Faculty Press.

———. 1986. Black American participation in American science: Winning some battles but losing the war. *Journal of Educational Equity and Leadership* 6 (1): 45–59.

———. 1987. The flow of black scientific talent: Leaks in the pipeline. *Humboldt Journal of Social Relations* 14 (1–2): 44–60.

———. 1991. Black participation and performance in science, mathematics, and technical education. In C. V. Willie, A. M. Garibaldi, and W. L. Reed. *The Education of African-Americans,* 123–128. New York: Auburn House.

———. 1998. The career patterns of African American Ph.D. chemists. Final Report submitted to National Science Foundation (SBR9222546). Arlington, Va.: National Science Foundation.

Pearson, W., Jr., and H. K. Bechtel. 1989. *Blacks, Science, and American Education.* New Brunswick, N.J.: Rutgers University Press.

Pearson, W., Jr., C. Ness, and E. Hoban. 1999. Race, gender, and the baccalaureate origins of Ph.D. chemists. *Journal of Women and Minorities in Science and Engineering* 5 (2): 97–112.

Perkins, L. M. 1997. The African American female elite: The early history of African American women in the Seven Sister colleges, 1880–1960. *Harvard Educational Review* 67 (4): 718–756.

Persell, C. H. 1977. *Education and Inequality: A Theoretical and Empirical Synthesis.* New York: Free Press.

Peter, K., and L. Horn. 2005. Gender differences in participation and completion of undergraduate education and how they have changed over time. Postsecondary Education Descriptive Analysis Reports, NCES 2005:169. Washington, D.C.: U.S. Department of Education.

Powell, R. R., and J. Garcia. 1985. The portrayal of minorities and women in selected elementary science series. *Journal of Research in Science Teaching* 22 (6): 519–533.

Project Kaleidoscope. 1991. *What Works: Building Natural Science Communities.* Washington, D.C.: Project Kaleidoscope.

Putnam, R. D. 2000. *Bowling Alone: The Collapse and Revival of American Community.* New York: Simon and Schuster.

Ramirez, F. O., and C. M. Wotipka. 2001. Slowly but surely? The global expansion of women's participation in science and engineering fields of study. *Sociology of Education* 74 (3): 231–251.

Rani, G. 2000. Measuring change in students' attitudes toward science over time: An application of latent variable growth modeling. *Journal of Science Education and Technology* 9 (3): 213–225.

Rayman, P., and B. Brett. 1995. Women science majors: What makes a difference in persistence after graduation? *The Journal of Higher Education* 66 (4): 388–414.

———. 1993. *Pathways for Women in the Science: The Wellesley Report, Part I.* Wellesley, Mass.: Wellesley College, Center for Research on Women.

Reinharz, S. 1992. *Feminist Methods in Social Research.* New York: Oxford University Press.

Rice, J. K., and A. Hemmings. 1988. Women's colleges and women achievers: An update. *Signs* 13 (3): 546–559.

Rosamond, F. A. N. 1991. A century of women's participation in the MAA and other organizations. In P. Kenschaft, ed., *Winning Women into Mathematics,* 31–46. Washington, D.C.: Mathematical Association of America.

Roschelle, A. R. 1999. Gender, family structure, and social structure: Racial ethnic families in the United States. In M. M. Ferree, J. Lorber, and B. B. Hess, eds., *Revisioning Gender,* 311–340. Thousand Oaks, Calif.: Sage.

Rosenthal, J. W. 1996. *Teaching Science to Language Minority Students: Theory and Practice.* Philadelphia: Multilingual Matters Ltd.

Rossiter, M. S. 1982. *Women Scientists in America: Struggles and Strategies to 1940.* Baltimore: Johns Hopkins University Press.

Rothenberg, P. S. 1992. *Race, Class, and Gender in the United States.* New York: St. Martins.

Rubenfeld, M. F., and F. D. Gilroy. 1991. Relationship between college women's occupation interests and a single-sex environment. *The Career Development Quarterly* 40 (1): 64–71.

Ryan, A. M. 2000. Peer groups as a context for socialization of adolescents' motivation, engagement, and achievement in school. *Educational Psychologist* 35: 101–111.

Sadowski, M. 2003. Why identity matters at school. In M. Sadowski, ed., *Adolescents at School: Perspectives on Youth, Identity, and Education,* 1–5. Boston: Harvard Education Press.

Sage, G. H. 1990. *Power and Ideology in American Sport: A Critical Perspective.* Champaign, Ill.: Human Kinetics Books.

Sampson, R. J., J. D. Morenoff, and G. Thomas. 2002. Assessing "neighborhood effects": Social processes and new directions in research. *Annual Review of Sociology* 28: 443–478.

Sanders, J. G. 1997. Overcoming obstacles: Academic achievement as a response to racism and discrimination. *Journal of Negro Education* 66 (1): 83–93.

Scanzoni, J. 1975. Sex roles, economic factors, and marital solidarity in black and white marriage. *Journal of Marriage and the Family* 37 (1): 130–144.

Scantlebury, K., T. Tal, and J. Rahm. 2006. "That don't look like me": Stereotypic images of science: Where do they come from and what can we do with them? *Cultural Studies of Science Education* 1 (3). http://www.springer link.com/content/wm8g5g635k251r04 (accessed July 8, 2008).

Schwager, S. 1987. Educating women in America. *Signs* 12 (2): 333–372.

Sewell, T. 2000. Beyond institutional racism: Tackling the real problems of black underachievement. *Multicultural Teaching (MCT)* 18 (2): 27–33.

Sewell, W. H., and R. H. Hauser. 1972. Causes and consequences of higher education: Models of the status attainment process. *American Journal of Agricultural Economics* 54 (5): 851–861.

Seymour, E., and N. M. Hewitt. 1997. *Talking about Leaving: Why Undergraduates Leave the Sciences.* Boulder, Colo.: Westview Press.

Sherman, H. A., and E. Fennema. 1977. The study of mathematics by high school girls and boys: Related variables. *American Educational Research Journal* 14 (2): 159–168.

Simpson, R. D., and J. S. Oliver. 1990. A summary of major influences on attitude toward and achievement in science among adolescent students. *Science Education* 74 (1): 1–18.

Smedley, A. 2002. Science and the idea of race: A brief history. In J. M. Fish, ed., *Race and Intelligence: Separating Science from Myth,* 145–176. Mahwah, N.J.: Lawrence Erlbaum Associates.

Spain, D., and S. M. Bianchi. 1996. *Balancing Act: Motherhood, Marriage, and Employment among American Women.* New York: Russell Sage Foundation.

Stake, J. E. 2003. Understanding male bias against girls and women in science. *Journal of Applied Social Psychology* 33: 1–17.

Stake, J. E., and K. R. Mares. 2001. Science enrichment programs for gifted high school girls and boys: Predictors of program impact on science confidence and motivation. *Journal of Research in Science Teaching* 38 (10): 1065–1088.

Stake, J. E., and S. D. Nickens. 2005. Adolescent girls' and boys' science peer relationships and perceptions of the possible self as scientist. *Sex Roles* 52 (1–2): 1–11.

Steele, C. M. 1997. A threat in the air: How stereotypes shape intellectual identity and performance. *American Psychologist* 52 (6): 613–629.

Stempel, L., C. Ney, and J. Ross J. 2001. Institutionalizing change. In C. Ney, J. Ross, and L. Stempel, eds., *Flickering Clusters: Women, Science, and Collaborative Transformations,* 96–151. Madison: University of Wisconsin Press.

Summers, L. H. 2005. Remarks at NBER Conference on Diversifying the Science and Engineering Workforce. Cambridge, Mass.: President and Fellows of Harvard College. http://www.president.harvard.edu/speeches/2005/nber.html (accessed October 10, 2007).

Swarat, S., D. Drane, H. D. Smith, G. Light, and L. Pinto. 2004. Opening the gateway: Increasing minority student retention in introductory science courses. *Journal of College Science Teaching* (September): 18–23.

Talton, E. L., and R. D. Simpson. 1985. Relationships between peer and individual attitudes toward science among adolescent students. *Science Education* 69 (1): 19–24.

Terry, J. J., and W. E. Baird. 1997. What factors affect attitudes toward women in science held by high school biology students? *School Science and Mathematics* 97 (2): 78–86.

Thomas, G. E. 1997. Race relations and campus climate for minority students: Implications for higher education desegregation. In C. Herring, ed., *African Americans and the Public Agenda: The Paradoxes of Public Policy,* 171–189. Thousand Oaks, Calif.: Sage.

Thomas, V. L. 1989. Black women engineers and technologists. *SAGE: A Scholarly Journal on Black Women* 6 (2): 24–32.

Thompson, L., and A. J. Walker. 1995. The place of feminism in family studies. *Journal of Marriage and the Family* 57 (4): 847–865.

Thornbery, O. T., and J. T. Massey. 1988. Trends in United States telephone coverage across time and subgroups. In R. Groves et al., eds. *Telephone Survey Methodology,* 25–49. New York: Wiley & Sons.

Tidball, E. M. 1975. Women on campus and you. *Liberal Education* 61 (2): 285–292.

———. 1980. Women's colleges and women achievers revisited. *Signs* 5 (3): 504–517.

———. 1986. Baccalaureate origins of recent natural science doctorates. *Journal of Higher Education* 57 (6): 448–454.

Tuch, S., L. Sigelman, and J. MacDonald. 1999. Trends: Race relations and American youth. *Public Opinion Quarterly* 63 (1): 109–148.

U.S. Bureau of the Census. 1995. *New Comprehensive African American Report.* Accessed at http://www.umsl.edu/-libweb/blackstudies/afcensus.htm (accessed June 12, 2006).

———. 2000. Census 2000 Summary File (SF3). http://factfinder.census.gov (accessed July 11, 2008).

Vinchez-Gonzalez, J. M., and F. J. P. Palacios. 2006. Image of science in cartoons and its relationiship with the image in comics. *Physics Education* 41 (3): 240–249.

Vining Brown, S. 1994. Minority women in science and engineering education: Final report. Princeton, N.J.: Educational Testing Service.

Von Sentima, I. 1985. *Blacks in Science: Ancient and Modern.* New Brunswick, N.J.: Transition Books.

Wade, B. H. 1993. The gender role and contraceptive attitudes of young men: Implications for future African American families. *Urban League Review* 16 (2): 57–65.

Walberg, H. 1984. Improving the productivity of America's schools. *Educational Leadership* 41 (8): 19–27.

Ware, N. C., and V. E. Lee. 1988. Sex differences in choice of college majors. *American Educational Research Journal* 25 (4): 593–614.

Warren, M. R., J. P. Thompson, and S. Saegert. 2001. The role of social capital in combating poverty. In S. Saegert, J. P. Thompson, and M. R. Warren, eds., *Social Capital and Poor Communities,* 1–28. New York: Russel Sage Foundation.

Wenner, G. 2003. Comparing poor, minority elementary students' interest and background in science with that of their white, affluent peers. *Urban Education* 38 (2): 153–172.

West, C., and S. Fenstermaker. 1995. Doing difference. *Gender and Society* 9 (1): 8–37.

Wilcox, C. 1989. Race, gender role attitudes, and support for feminism. *The Western Political Quarterly* 43 (1): 113–121.

Wilson, J., and M. Musick. 1997. The contribution of social resources to volunteering. *Social Science Quarterly* 79 (4): 799–814.

Wilson, K. L., and J. P. Boldizar. 1990. Gender segregation in higher education: Effects of aspirations, mathematics, achievements, and income. *Sociology of Education* 63 (1): 62–74.

Wilson-Sadbury, K. R. 1991. Resilience and persistence of African American males in post-secondary enrollment. *Education and Urban Society* 24 (1): 87–102.

Zinn, M. B., and D. T. Dill. 1996. Theorizing difference from multiracial feminism. *Feminist Studies* 22 (2): 321–331.

Zweigenhaft, R. L., and G. W. Domhoff. 1998. *Diversity in the Power Elite: Have Women and Minorities Reached the Top?* New Haven: Yale University Press.

Index

Sandra L. Hanson is Professor of Sociology and Research Associate at Life Cycle Institute, Catholic University. She is the author of *Lost Talent: Women in the Sciences* (Temple).